IARC MONOGRAPHS

ON THE

EVALUATION OF THE CARCINOGENIC

RISK OF CHEMICALS TO MAN:

Some aromatic amines,
hydrazine and related substances,
N-nitroso compounds and
miscellaneous alkylating agents

Volume 4

This publication represents the views
of the IARC Working Group on the Evaluation
of the Carcinogenic Risk of Chemicals to Man
which met in Lyon, 18-25 June 1973

IARC WORKING GROUP ON THE EVALUATION OF THE CARCINOGENIC RISK OF CHEMICALS
TO MAN: SOME AROMATIC AMINES, HYDRAZINE AND RELATED SUBSTANCES, N-NITROSO
COMPOUNDS AND MISCELLANEOUS ALKYLATING AGENTS

Lyon, 18-25 June 1973

Members[1]

Dr R.R. Bates, Chief, Experimental Pathology Branch, Carcinogenesis,
Division of Cancer Cause & Prevention, National Cancer Institute,
Bethesda, Maryland 20014, USA

Professor E. Boyland, TUC Institute of Occupational Health, London School
of Hygiene and Tropical Medicine, Keppel Street, London WC1, UK

Dr D.B. Clayson[2], The Eppley Institute for Research on Cancer, University
of Nebraska Medical Center, 42nd & Dewey Avenue, Omaha, Nebraska
68105, USA (Chairman)

Dr T.A. Gough, Research Division, Department of Trade and Industry,
Cornwall House, Stamford Street, London SE1 9NQ, UK

Dr A.E. Martin, Department of Health and Social Security, Alexander
Fleming House, Elephant and Castle, London SE1 6BY, UK

Dr H.G. Parkes, British Rubber Manufacturers Association Ltd, Health
Research Unit, Scala House, Holloway Circus, Birmingham B1ER, UK

Dr R. Preussmann, Deutsches Krebsforschungszentrum, Institut für Toxikologie und Chemotherapie, D 6900 Heidelberg 1, Kirschnerstrasse 6,
Postfach 449, Federal Republic of Germany (Vice Chairman)

Dr B. Teichmann, Zentralinstitut für Krebsforschung, Academie der
Wissenschaften der DDR, DDR-1115 Berlin-Buch, Lindenberger Weg 80,
German Democratic Republic

Professor R. Truhaut, Directeur, Centre de Recherches Toxicologiques,
Faculté des Sciences, Pharmaceutiques et Biologiques, Université
René Descartes, 4 avenue de l'Observatoire, 75005 Paris, France

Dr J.H. Weisburger, American Health Foundation, 2 East End Avenue, New
York, NY 10021, USA

---

[1] Unable to attend: Dr F.K. Dzhioev, Petrov Research Institute of
Oncology, Pesochny-2, Leningrad, USSR

[2] On sabbatical leave during 1973 from the Department of Experimental
Pathology and Cancer Research, University of Leeds, 171 Woodhouse Lane,
Leeds LS2 3AR, UK

Representative of the National Cancer Institute

Dr J. Cooper, Office of the Associate Director, Carcinogenesis, National Cancer Institute, National Institutes of Health, Bethesda, Maryland 20014, USA

Invited Guests

Dr P.S. Elias, Principal Medical Officer, Department of Health and Social Security, Alexander Fleming House, Elephant and Castle, London SE1 6BY, UK

Dr R. Kroes, Head of the Department of Oncology, Rijks Instituut voor de Volksgezondheid, Postbus 1, Bilthoven, The Netherlands

Dr K.E. McCaleb, Manager, Environmental Studies, Chemical Information Services, Stanford Research Institute, Menlo Park, California 94025, USA

Secretariat

Dr C. Agthe, Unit of Chemical Carcinogenesis (Secretary)

Dr H. Bartsch, Unit of Chemical Carcinogenesis

Dr N. Breslow, Unit of Epidemiology and Biostatistics

Dr P. Bogovski, Chief, Unit of Environmental Carcinogens

Dr M. El Batawi, Chief, Occupational Health Unit, WHO

Dr R. Montesano, Unit of Chemical Carcinogenesis

Mrs C. Partensky, Library Officer

Dr L. Tomatis, Chief, Unit of Chemical Carcinogenesis

Dr A.J. Tuyns, Unit of Epidemiology and Biostatistics

Mr E.A. Walker, Unit of Environmental Carcinogens

Mr J.D. Wilbourn, Unit of Chemical Carcinogenesis

CONTENTS

BACKGROUND AND PURPOSE OF THE IARC PROGRAMME ON THE EVALUATION OF CARCINOGENIC RISK OF CHEMICALS TO MAN

The International Agency for Research on Cancer (IARC) has initiated a programme on the evaluation of carcinogenic risk of chemicals to man. This programme was supported by a Resolution of the Governing Council at its Ninth Session concerning the role of IARC in providing government authorities with expert, independent scientific opinion on environmental carcinogenesis. As one means to this end, the Governing Council recommended that the Agency should continue to prepare monographs on the carcinogenic risk of individual chemicals to man.

In view of the importance of this programme and in order to expedite the production on monographs, the National Cancer Institute of the United States has provided IARC with additional funds for this programme.

The objective of this programme is to achieve and publish a balanced evaluation of data through the deliberations of an international group of experts in chemical carcinogenesis and to put into perspective the present state of knowledge with the final aim of evaluating the data in terms of possible human risk, as well as to indicate the need for research efforts to close the gaps in our knowledge.

SCOPE OF THE MONOGRAPHS

The monographs summarize the evidence for the carcinogenicity of individual chemicals in a condensed uniform manner for easy comparison. The data are compiled, reviewed and evaluated by a Working Group of experts. No recommendations are given concerning preventive measures or legislation, since these matters depend on risk-benefit evaluation, which seems best made by individual governments and/or international agencies such as WHO and ILO.

The first volume, published in 1972[1], covers a number of substances

---

[1] International Agency for Research on Cancer (1972) IARC Monographs on the Evaluation of Carcinogenic Risk of Chemicals to Man, 1, Lyon

not belonging to a particular chemical group; the second volume[1] contains monographs on some inorganic and organometallic compounds; and the third volume[2] covers a number of polycyclic aromatic hydrocarbons and heterocyclic compounds. The present volume is devoted to some aromatic amines, hydrazine and related substances, N-nitroso compounds and miscellaneous alkylating agents.

As new data on chemicals for which monographs have already been written and new principles for evaluation become available, re-evaluations will be made at future meetings, and revised monographs will be published as necessary. The monographs are being distributed to international and governmental agencies and will be available to industries and scientists dealing with these chemicals. They also form the basis of advice from IARC on carcinogenesis from these substances.

MECHANISM FOR PRODUCING THE MONOGRAPHS

As a first step, a list of chemicals for possible consideration by the Working Group is established. IARC then collects pertinent references regarding physico-chemical characteristics, production and use[3], occurrence and analysis, and biological data on these compounds. The material is summarized by an expert consultant or an IARC staff member, who prepares the first draft monographs, which in some cases is sent to another expert for comments. The drafts are circulated to all members of the Working Group about one month before the meeting, at which further additions to and deletions from the data are agreed upon and a final version of comments and evaluation on each compound is adopted.

---

[1] International Agency for Research on Cancer (1973) IARC Monographs on the Evaluation of Carcinogenic Risk of Chemicals to Man, 2, Some Inorganic and Organometallic Compounds, Lyon

[2] International Agency for Research on Cancer (1973) IARC Monographs on the Evaluation of Carcinogenic Risk of Chemicals to Man, 3, Some Polycyclic Aromatic Hydrocarbons and Heterocyclic Compounds, Lyon

[3] Data from Chemical Information Services, Stanford Research Institute, USA

Priority for the Preparation of Monographs

Priority for consideration is given mainly to chemicals for which some adequate experimental evidence of carcinogenicity exists and/or for which there is evidence of human exposure. However, neither human exposure nor potential carcinogenicity can be judged until all the relevant data have been collected and examined in detail. The inclusion of a particular compound in a monograph does not necessarily mean that the substance is considered to be carcinogenic. Equally, the fact that a substance has not yet been considered does not imply that it is non-carcinogenic.

Data on which the Evaluation was Based

With regard to the biological data, only published articles or papers already accepted for publication are reviewed. Every effort is made to cover the whole literature, but some studies may have been inadvertently overlooked. Since the monographs are not intended to be a review of the literature, they contain only data considered relevant by the Committee, and it is not possible for the reader to determine whether a certain study was considered or not. It is therefore important that research workers who are aware of important data which may change the evaluation make them available to the Unit of Chemical Carcinogenesis of the International Agency for Research on Cancer, Lyon, France, in order that they can be considered for a possible re-evaluation.

The Working Group

The members of the Working Group who participated in the consideration of particular substances are listed at the beginning of each publication. Each monograph bears a footnote indicating the date of the meeting at which it was considered. The members of the Working Group are invited by IARC to serve in their individual capacities as scientists and not as representatives of their governments or of any institute with which they are affiliated.

GENERAL REMARKS ON THE EVALUATION

Terminology

The term "chemical carcinogenesis" in its widely accepted sense is used to indicate the induction or enhancement of neoplasia by chemicals. It is recognized that, in the strict etymological sense, this term means the induction of cancer. However, common usage has led to its employment to denote the induction of various types of neoplasm. The terms "tumourigen", "oncogen" and "blastomogen" have all been used synonymously with "carcinogen", although occasionally "tumourigen" has been used specifically to denote the induction of benign tumours.

Response to Carcinogens

For present practical purposes, in general, no distinction is made between the induction of tumours and the enhancement of tumour incidence, although it is noted that there may be fundamental differences in mechanisms that will eventually be elucidated.

The response in experimental animals to a carcinogen may be observed in several forms:

(a) as a significant increase in the frequency of one or several types of neoplasm, as compared with the control;

(b) as the occurrence of neoplasms not observed in control animals;

(c) as a decreased latent period as compared with control animals;

(d) as a combination of (a) and (c).

Qualitative Aspects

The qualitative nature of neoplasia has been much discussed. Many instances of carcinogenesis involve the induction of both benign and malignant tumours. There are few, if any, recorded instances in which only benign tumours are induced; their occurrence in experimental systems indicates that the same treatment may increase the risk of malignant tumours also.

In experimental carcinogenesis, the type of cancer seen is often the same as that recorded in human studies (e.g., bladder cancer in man, monkeys, dogs and hamsters after administration of 2-naphthylamine). In other instances, however, a chemical will induce different neoplasms or neoplasms at different sites in different animal species (e.g., benzidine, which induces hepatic carcinoma in the rat, but bladder carcinoma in man).

Purity of the Compound Tested

In any evaluation of biological data with respect to a possible carcinogenic risk, serious consideration must be given to the purity of the chemicals tested. Frequently, in older studies and sometimes in more recent ones as well, insufficient attention was given to the purity of materials synthesized or to their stability under conditions of storage or administration. Thus, lesions described in such studies may have resulted from impurities rather than from the chemical described, or negative results may have resulted from degradation of the test chemical. These facts may account for some inconsistencies in the reported literature. Further information on stability is documented in the individual monographs.

In this series of monographs these considerations are particularly important in the case of commercial 1-naphthylamine, which may contain unknown amounts of 2-naphthylamine. The presence of bis(chloromethyl) ether as an impurity in chloromethyl methyl ether must also be taken into consideration in evaluating the carcinogenic risk of the latter compound.

Quantitative Aspects

Dose-response studies are important in the evaluation of human and animal carcinogenesis. Sometimes, the only way in which a causal effect can be established with confidence is by the observation of increased incidence of neoplasms over the control in relation to increased exposure. It is hoped that, eventually, dose-response data may be used for assessment in carcinogenesis in the same way that they are used in general toxicological practice.

Extrapolation from Animals to Man

No attempt has been made to interpret the animal data in the absence of human data in terms of possible human risk, and no distinction has been made between weak and strong carcinogens, since no objective criteria are at present available to do so. These monographs may be reviewed if some such criteria should be elaborated. In the meantime, the critical assessment of the validity of the animal data given should help national and/or international authorities to make decisions concerning preventive measures or legislation in the light of WHO recommendations on food additives[1], drugs[2] and occupational carcinogens[3].

Evidence of Carcinogenicity to Humans

Evidence that a particular chemical is carcinogenic in man depends on clinical and epidemiological data which may be in the main descriptive, retrospective or prospective. Descriptive studies may identify a cluster of or a change in rates for a particular neoplasm in a subgroup of the population which suggest the influence of carcinogens in the environment. Retrospective studies (i.e., case-control studies that go into the histories of persons with or without cancer) have revealed occupational carcinogens (e.g., shale oil, chromates, asbestos, β-naphthylamine, benzidine) or iatrogenic carcinogens (e.g., chlornaphazin, thorotrast, diethylstilboestrol).

Once a relationship is known or suspected between an exposure and cancer, prospective studies (i.e., follow-up or cohort studies of exposed and unexposed groups) will identify more precisely the magnitude of the risk and may clarify time relationships, dose-response effects and other aspects of cancer induction. Whenever possible the Working Group con-

---

[1] World Health Organization (1961) Fifth Report of the Joint FAO/WHO Expert Committee on Food Additives. Wld Hlth Org. techn. Rep. Ser., No. 220, pp. 5, 18 and 19

[2] World Health Organization (1969) Report of a WHO Scientific Group. Wld Hlth Org. techn. Rep. Ser., No. 426, pp. 19, 21 and 22

[3] World Health Organization (1964) Report of a WHO Expert Committee. Wld Hlth Org. techn. Rep. Ser., No. 276, pp. 29 and 30

sidered evidence of the influence of variables other than the agent under suspicion in inducing the cancer under study (e.g., cigarette-smoking in the study of lung cancer among asbestos workers).

Finally, if man does develop cancer from a specific chemical, its removal from the environment should be followed eventually by epidemiological evidence of a decline in the frequency of the neoplasm in the exposed group.

Mixtures and Groups of Carcinogens

Mixtures of chemicals are sometimes associated with the occurrence of cancers in man, but no information is available on the specific components. Continuing efforts should be made to elucidate the role of the various components and impurities in substances to assist in planning better preventive measures and to provide a basis for assessing similar hazards. There are situations where carcinogens may occur in groups in the human environment and where it is not yet possible to attribute the observed effects to individual substances. This is notably so in the case of the polycyclic aromatic hydrocarbons, heterocyclic compounds and certain aromatic amines.

EXPLANATORY NOTES ON THE MONOGRAPHS

In sections 1, 2 and 3 of each monograph, except for minor remarks, the data are recorded as given by the author, whereas the comments by the Working Group are given in section 4, headed "Comments on Data Reported and Evaluation".

Title of Monograph

The monograph has as its title the chemical name of the substance under consideration. For this name, the chemical abstract nomenclature is normally used. Other chemical synonyms, common names and trade names are indicated in this section, but no attempt was made to be all-inclusive.

Chemical and Physical Data (section 1)

Chemical and physical properties include data that might be relevant to carcinogenicity (for example, lipid solubility) and those that concern

identification. Where relevant, data on solubility, volatility and
stability are indicated. All data except those for "Technical products
and impurities" refer to the pure substances, unless otherwise indicated.

Production, Use, Occurrence and Analysis (section 2)

With regard to the data on use and occurrence of chemicals presented
in the monographs, IARC has collaborated with the Stanford Research
Institute, USA, with the support of the National Cancer Institute of the
United States, in order to obtain production figures of chemicals and
their pattern of use. These data more commonly refer to the United States
than to other countries solely as a result of the availability to the
Working Group of more data from that country than from others. It should
not be implied that that nation is the sole source or even the major source
of any individual chemical or that enquiries have been made in only one
country.

In some countries, there are also legal restrictions on the conditions
under which certain carcinogens and suspect chemicals can be handled.
Examples of these are given in section 2 (b) when such information was
available to the Working Group.

It is hoped that in future revisions of these monographs, more
information on use and legislation can be made available to IARC from
other countries.

Biological Data Relevant to the Evaluation of Carcinogenic Risk to Man (section 3)

As pointed out earlier in this introduction, the monographs are not
intended to itemize all studies reported in the literature. Although
every effort was made to review the whole literature, some studies were
purposely omitted (a) because of their inadequacy[1,2,3,4] (e.g., too short

---

[1] World Health Organization (1958) Second Report of the Joint FAO/WHO
Expert Committee on Food Additives. Wld Hlth Org. techn. Rep. Ser.,
No. 144

a duration, too few animals, poor survival or too small a dose), (b) because they only confirmed findings which had already been fully described or (c) because they were judged irrelevant for the purpose of the evaluation. However, in certain cases, reference is made to studies which did not meet established criteria of adequacy, particularly when this information was considered a useful supplement to other reports or when it may have been the only data available. This does not, however, imply acceptance of the adequacy of experimental design in these cases.

In general, the data recorded in this section are summarized as given by the author; however, certain shortcomings of reporting or of experimental design are also mentioned and minor comments by the Working Group are given in brackets. The essential critical comments by the Working Group are made in section 4 ("Comments on Data Reported and Evaluation").

Carcinogenicity and related studies in animals (3.1)

Mention is made of all routes of administration by which the compound has been tested and of all species in which relevant tests have been carried out. In some cases where similar results were obtained by several authors, reference is made to a summary article. Quantitative data are given in so far as they will enable the reader to realize the order or magnitude of the effective dose. In general the doses are indicated as

---

[2] World Health Organization (1961) Fifth Report of the Joint FAO/WHO Expert Committee on Food Additives. Wld Hlth Org. techn. Rep. Ser., No. 220

[3] World Health Organization (1967) Scientific Group. Wld Hlth Org. techn. Rep. Ser., No. 348

[4] UICC (1969) Carcinogenicity Testing. UICC techn. Rep. Ser., Vol. 2

they appear in the original paper; however, sometimes conversions have been made for better comparison, and these are given in brackets. Where other compounds were investigated simultaneously in the same experiment, results on these are also included, in order to compare their biological activity with that of the substance in question.

Other relevant biological data (3.2)

The data reported in this section are divided into the following sub-sections:

(a) information on the metabolic fate in animals including localization into tissues;

(b) similar information in man;

(c) comparison of the animal and human data;

(d) the carcinogenicity of known metabolites.

Data on toxicity are included occasionally, if considered relevant. In some monographs reference is made to malignant transformation of cultured cells exposed in vitro to the chemical. Such studies are only included when in vitro transformation was confirmed by in vivo transplantation.

Observations in man (3.3)

Epidemiological studies are summarized. This sub-section also includes, where relevant, summaries of reports of cases of cancer in man that have been related to possible exposure to the chemical.

Comments on Data Reported and Evaluation (section 4)

This section includes the critical view of the Working Group on the data reported. It is purposely kept as brief as possible since it should be read in conjunction with the data recorded.

Animal data (4.1)

The animal species mentioned are those in which the carcinogenicity of the substances was investigated adequately. In the case of inadequate studies, when mentioned, comments to that effect are sometimes included. The route of administration used in experimental animals that is similar

16

to the possible human exposure (ingestion, inhalation and skin exposure) is given particular mention. In some cases, selective tumour sites are also indicated. Comparison of the carcinogenic activity with that of other related substances is made in some cases when parallel experiments were run in the same study. If the substance has produced tumours on prenatal exposure or in single-dose experiments, this is also indicated. This subsection should be read in the light of comments made in the section "Extrapolation from animals to man".

Human data (4.2)

The significance of case reports and epidemiological studies is discussed and the data are interpreted in terms of possible human risk.

CARCINOGENICITY OF THE AROMATIC AMINES IN MAN

The epidemiology of aromatic amine cancer is essentially the epidemiology of human bladder cancer of industrial origin, and a large number of published reports has appeared in the scientific literature during the past 75 years which indicate both the prevalence and the widespread distribution of this condition in many manufacturing industries where these chemicals have been used. In the monographs which follow it is understood that the function and responsibility of the Working Group is to assess and evaluate only those epidemiological data that can be specifically related to the particular chemical compound under consideration. The rigid application of this principle must inevitably result in the exclusion from consideration of highly important and otherwise relevant epidemiological data which relate either to mixed exposures or to hypothetical exposures to those trace quantities of carcinogens known to be present as impurities in other compounds, many of which are not considered in the present monographs series.

For these reasons, it seemed desirable to review briefly here the epidemiological data in this wider context, so that the problems may be seen in clearer perspective.

Since Rehn (1895) first reported four cases of bladder cancer in a group of 45 men engaged in the manufacture of fuchsin, much time and research have been devoted to the identification of the chemical carcinogen(s) chiefly implicated. Although aniline, the toxic effects of which were already well known, was the first of the amines to come under suspicion, it can be stated with reasonable certainty that the dyestuffs workers studied by Rehn must have been exposed to a number of other chemical intermediates. Later investigations served largely to exonerate aniline, and attention soon became focussed upon benzidine and the naphthylamines. In 1921, an ILO report (ILO, 1921) listed a number of compounds then under suspicion amongst the aromatic amines but singled out 2-naphthylamine and benzidine for special mention.

During the next two decades, case reports indicating the hazard to chemical workers handling these compounds were accumulating in many different countries, and, for the first time, experimental evidence which served to establish the potent carcinogenicity of 2-naphthylamine and benzidine was beginning to emerge. Hueper et al.(1938) first induced bladder tumours in dogs with 2-naphthylamine, and confirmation of his findings was provided by the experiments of Bonser (1943) and Bonser et al.(1956). In 1950, Spitz et al. induced bladder tumours at 7, 8 and 9 years in 3 out of 7 dogs by the administration of benzidine (see also IARC, 1972).

Thus, by 1950 there existed a substantial amount of experimental and clinical evidence to confirm the occupational aetiology of bladder cancer in dyestuffs workers and also to identify the principal carcinogens. At this time the attention of epidemiologists was rightly concentrated upon the undeniably grave problems which confronted the chemical manufacturers of dyestuffs, and there was no real suspicion as yet that the carcinogenic hazard would prove to be more far reaching than was originally feared. It was the work of Robert Case and his colleagues in the UK which revealed the involvement of hitherto unsuspected areas of industry. Case began his classic studies of the British dyestuffs industry in 1948, undertaking an extensive field study designed to establish not only the magnitude of the problem but also which of the suspect carcinogens were chiefly implicated.

18

Case's study was one of the most painstaking and exhaustive epidemiological enquiries ever carried out and inevitably took some time to complete. However, by 1952 he had been able to locate 455 cases of bladder tumour in the British chemical industry, and 311 of these could be shown to have occurred in men known to have worked with the suspect aromatic amines, benzidine, 1-naphthylamine, 2-naphthylamine, auramine and magenta. Case further demonstrated that the chance of a 2-naphthylamine worker developing a bladder cancer was 61 times that found in the general population, that of a benzidine worker 19 times, and that of a worker exposed to commercial 1-naphthylamine 16 times. In addition to noting the increased incidence of bladder cancer, Case made the important observation that the occupational cancers were occurring at a much earlier age than were the non-occupational tumours. He established also that the latent period for tumour induction averaged 16 years for 2-naphthylamine and benzidine workers. Williams (1958), reporting on a study of men drawn from the same population studied by Case, gave striking evidence of the effects of severe exposure conditions. He described a group of 78 men engaged in distilling 2-naphthylamine and benzidine under poor conditions, at a time when the hazard was not well recognized. Amongst these he found 35 men (45%) who had developed bladder tumours, but the incidence rose steeply with increasing exposure times. Among 43 men with less than 3 years' exposure there were only 7 cases (16%), among 17 men with 3-5 years' exposure there were 11 cases (65%) and among 18 men with 5 or more years' exposure there were no fewer than 17 cases (94%).

It is, however, an almost coincidental observation made during the course of his investigation which led Case to what was possibly his most significant finding. Whilst examining regional mortality data, Case noted an apparently high prevalence of bladder cancer amongst a group of men all of whom could be identified as having worked in one large rubber factory. For the years 1936-1950, when four cases would have been predicted, 22 cases were actually found. At the same time enquiries revealed that rubber antioxidants then in use were manufactured from 1- and 2-naphthylamine, and that men involved in the manufacturing process were also suffering from bladder cancer. Not unnaturally, therefore, Case extended his enquiries into the rubber industry and produced in 1954 a report which established that an enhanced risk of bladder cancer was affecting British male rubber workers in the period 1936-1951. Since that time, further evidence has come to light to confirm Case's observations, but the precise identification of the active carcinogen has remained in some doubt (Veys, 1969). Whilst it is known that some rubber antioxidants used prior to 1949 were manufactured from 1- and 2-naphthylamine, the finished products contained no more than 0.25% of unreacted 2-naphthylamine, and it has been suggested that this level of impurity could scarcely account for the severity of the hazard experience. An alternative theory is that the final product, the precise composition of which has never been determined, was carcinogenic per se; but although this suggestion has gained significant support amongst informed scientists it was firmly rejected in a legal test case fought in the English High Court in 1971. Whatever the final verdict upon the question may be, the significance of Case's discovery is increasingly apparent. The occupational hazard in the chemical industry may be manifestly severe for workers who have been subjected for prolonged periods to high levels of exposure.

By contrast, exposure during the use of these compounds in the rubber industry has been of a different character. The intensity of the hazard, presumably resulting from small impurities of 2-naphthylamine, has been relatively low. Veys (1972) shows that the overall incidence of bladder cancer found in his population of rubber workers was four times that of the general population, so that the level of risk to the rubber worker is correspondingly reduced. The dispersion of the hazardous materials was, however, immeasurably greater than in the chemical industry, since carcinogenic antioxidant powders, widely distributed to rubber manufacturers throughout the country, were incorporated into the rubber compound in the earliest stages of its manufacture and persisted right through to the finished product. It is, of course, important to understand that in the compounding, masticating, extruding and vulcanizing stages of production the rubber compound is subjected to high temperature treatment which creates the opportunity for significant exposures to any vapour hazard. There is epidemiological evidence to show that in the use of carcinogenic antioxidants the workers chiefly at risk were those involved either in the compound mixing departments or in the final vulcanizing operation. At these lower levels of exposure there was, therefore, the opportunity for many more men to be brought into contact with the hazard, and it now seems probable that the resulting number of cases of occupational bladder cancer will be no fewer than those to be found in the chemical industry. Nor is this spread confined to the rubber industry alone. Scott (1962), in his monograph, indicated the probable extension of the hazard to cable manufacturers (who have used identical antioxidants), the makers of pigments and paints (users of similar intermediates), workers in the paper industry (users of auramine and magenta) and of course large numbers of scientific and medical laboratory personnel (users of benzidine). Unfortunately, however, no epidemiological data relating to these occupations, other than one study by Davies (1965) of the cable industry, appear to be available at present.

Although the manufacture of 2-naphthylamine and the use of suspect rubber antioxidants with significant free amine impurities were abandoned by many countries in 1950, or even earlier, it has to be noted that in some countries manufacture and use have continued at least into the 1970's, and the conclusion that many cases of industrial bladder cancer must remain to be revealed is inescapable. It is evident from studies such as those of Goldwater et al.(1965) that the British experience of a major bladder cancer hazard extending far beyond the confines of the chemical industry is not unique, and there would appear to be a real need for further epidemiological study of this problem in other parts of the world. There are, however, a number of questions which are as yet unresolved.

The weight of scientific and epidemiological evidence to incriminate 2-naphthylamine and benzidine is certainly overwhelming, but this relates to the environmental presence of these compounds in what must at present, and for want of more precise experimental evidence, be described as probably carcinogenic quantities. The antioxidants which proved to be carcinogenic in the rubber industry can be shown to have contained some 2,500 ppm of free 2-naphthylamine; when these were withdrawn they were replaced by others which were then believed to be entirely free from impurity. The most important of these, and deserving special mention, is N'-phenyl-2-naphthylamine, which should not be confused with the highly carcinogenic 2-naphthylamine. Since it is made from 2-naphthol and aniline, it was not to be expected that N'-phenyl-2-naphthylamine would contain any measurable quantities of 2-naphthylamine, and only the more sophisticated analytical techniques now available reveal that up to 50-100 ppm of that substance have probably been present in the commercial grades. The Working Group is aware that epidemiological studies are now proceeding which are designed to reveal any carcinogenic effects resulting from industrial exposure to commercial N'-phenyl-2-naphthylamine.

References

Bonser, G.M. (1943) Epithelial tumours of the bladder in dogs induced by pure β-naphthylamine. J. Path. Bact., 55, 1

Bonser, G.M., Clayson, D.B., Jull, J.W. & Pyrah, L.N. (1956) The carcinogenic activity of 2-naphthylamine. Brit. J. Cancer, 10, 533

Case, R.A.M., Hosker, M.E., McDonald, D.B. & Pearson, J.T. (1954) Tumours of the urinary bladder in workmen engaged in the manufacture and use of certain dyestuff intermediates in the British chemical industry. I. The role of aniline, benzidine, alpha-naphthylamine and beta-naphthylamine. Brit. J. industr. Med., 11, 75

Davies, J.M. (1965) Bladder tumours in the electric cable industry. Lancet, ii, 143

Goldwater, L.J., Rosso, A.J. & Kleinfeld, M. (1965) Bladder tumors in a coal-tar dye plant. Arch. environm. Hlth., 11, 814

Hueper, W.C., Wiley, F.H. & Wolfe, H.D. (1938) Experimental production of bladder tumors in dogs by administration of beta-naphthylamine. J. industr. Hyg., 20, 46

International Agency for Research on Cancer (1972) IARC Monographs on the Evaluation of Carcinogenic Risk of Chemicals to Man, 1, p.50

International Labour Office (1921) Studies and Reports, Series F, No.1, p.6

Rehn, L. (1895) Blasengeschwülste bei Fuchsin-arbeitern. Arch. klin. Chir., 50, 588

Scott, T.S. (1962) Carcinogenic and chronic toxic hazards of aromatic amines, Amsterdam, New York, Elsevier

Spitz, S., Maguigan, W.H. & Dobriner, K. (1950) The carcinogenic action of benzidine. Cancer (Philad.), 3, 789

Veys, C.A. (1969) Two epidemiological enquiries into the incidence of bladder tumours in industrial workers. J. nat. Cancer Inst., 43, 219

Williams, M.H.C. (1958) Occupational tumours of the bladder. In: Raven, R.W., ed., Cancer, Vol.3, London, Butterworths, p.337

i

AROMATIC AMINES

AND

RELATED COMPOUNDS

ANILINE*

1. Chemical and Physical Data

1.1 Synonyms and trade names

Chem. Abstr. No.: 62-53-3

Aminobenzene; aminophen; aniline oil; benzenamine; blue oil;
C.I. Oxidation Base 1 (Colour Index); C.I. 76000 (Colour Index);
phenylamine; appears in early literature as "krystallin", "cyanol",
"kyanol" and "benzidam"

1.2 Chemical formula and molecular weight

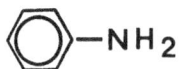 $-NH_2$ C_6H_7N Mol. wt: 93.1

1.3 Chemical and physical properties of the pure substance

(a) Description: An oily liquid; colourless with a bluish
    fluorescence when freshly distilled. Darkens on exposure to
    air and light and polymerizes to a resinous mass. Has a
    characteristic aromatic odour and a burning taste

(b) Boiling-point: $184^{O}C$

(c) Melting-point: $-6.2^{O}C$

(d) Specific gravity: d_{20}^{20} 1.022

(e) Refractive index: d_{D}^{20} 1.5863

(f) Volatility: Volatile in steam; vapour pressure is 0.67 mm Hg
    at $25^{O}C$, and 10 mm Hg at $69.4^{O}C$

---

* Considered by the Working Group in Lyon, June 1973.

(g) Solubility: Miscible with ethanol, ether, benzene and probably with most organic solvents and lipids; soluble in dilute hydrochloric acid. At 25°C water dissolves 3.7% of aniline and aniline dissolves 5.3% of water. One gram of aniline will dissolve in 15 ml of boiling water.

(h) Chemical reactivity: It has the general characteristics of primary aromatic amines. It is a weak base (pK$_b$ 9.3) and forms crystalline salts with strong acids. It forms a diazonium ion with nitrous acid, which couples readily with phenols and aromatic amines. It is easily acylated and alkylated. The amino group reacts readily with aldehydes, and the ring is easily substituted (e.g., by chlorine). Aniline is easily oxidized to complex self-condensation dyes known as the aniline blacks.

1.4 Technical products and impurities

Available in commercial and reagent grades. Typical commercial grade contains 99.9% (min.) aniline, 0.0002% (max.) nitrobenzene and 0.1% (max.) moisture[1]. These products generally contain low levels of oxidation products and may contain other aromatic amines if prepared from impure starting materials. Pre-1900 samples were alleged to have also contained 4-aminobiphenyl as an impurity (Walpole et al., 1952).

2. Production, Use, Occurrence and Analysis

One review concerning aniline and its derivatives has been published (Kouris & Northcott, 1963).

(a) Production and use[1]

Aniline was first produced in 1826 by the dry distillation of indigo. It was isolated from coal-tar in 1834, first synthesized in 1842, and was first manufactured commercially in 1847 (Kouris & Northcott, 1963). Aniline can also be produced by the reduction of nitrobenzene, the amination of chlorobenzene, and the amination of phenol.

---

[1] Data from Chemical Information Services, Stanford Research Institute, USA.

One plant in Japan was reported to be making aniline by the amination of phenol in 1971 (Anon., 1971a), but in the United States the reduction of nitrobenzene is the only method used for commercial production. The total US production (by seven producers) was reported to have been 166 million kg in 1971 (US Tariff Commission, March 1973), and preliminary data indicate that the total US production of aniline was 176 million kg in 1972 (US Tariff Commission, March 14, 1973). In 1970, US imports through the principal customs districts were reported to have been 0.53 million kg, all from the United Kingdom (US Tariff Commission, August 1971).

In 1968, the Federal Republic of Germany was reported to have had three producers of aniline with a total production of "aniline and salts" of 60 million kg in 1967 (US Department of Commerce, September 1970). In 1969, one of the companies was reported to have increased its annual capacity to 65 million kg, and a new plant with a capacity of 25 million kg was scheduled to be opened in 1972 (Anon., 1969).

In 1972, the production capacity of the five aniline producers in Japan was estimated to have been 68 million kg per year (Anon., 1972), but the production level is believed to have been considerably below this. In 1971, the capacity of the only French aniline plant was reported to have been 22 million kg per year (Anon., 1971b), but no information is available on production quantitites. In 1971, India's first aniline plant, with a capacity of 6 million kg per year, was reported to have started production (Anon., 1971d). In 1967, an aniline plant with an annual capacity of 0.57 million kg per year was reported to be under construction in the Arab Republic of Egypt (US Department of Commerce, March 1967).

In 1969, Italy was reported to have one producer of aniline, and the United Kingdom was reported to have six producers in 1970. It is known than aniline is also produced in the German Democratic Republic, the Union of Soviet Socialist Republics and the People's Republic of China, but no other information is available.

Consumption patterns are not available for the other major industrial countries, but these are believed to be similar to that in the US where nearly all aniline is used as an intermediate in the production of other chemicals. In 1971, one source estimated the following US consumption pattern for aniline: rubber-processing chemicals, 55%; isocyanates, 20%; dyestuffs, 10%; other uses (e.g., the manufacture of photographic chemicals, pharmaceuticals and agricultural chemicals), 15% (Anon., 1971c). It is believed that the consumption pattern is changing rapidly at this time because of the increasing use of aniline for isocyanate production.

The production of thiazole accelerators (e.g., 2-mercaptobenzothiazole and derivatives) is estimated to have required 14 million kg of aniline in the US in 1970. In 1970, a total of 14 million kg of aniline was also used in the production of diphenylamine, which was used either directly in rubber-compounding or as an intermediate in the production of other rubber-processing chemicals. It is estimated that a further 8 million kg of aniline were used in 1970 to make technical grade hydroquinone, the latter being used either directly in rubber-compounding or as an intermediate in the production of other rubber antioxidants.

Aniline is used as the starting material in the production of 4,4'-methylenediphenyl isocyanate and polymethylene polyphenylisocyanate. Combined US production of these isocyanates is estimated to have been 76 million kg in 1970. It is anticipated that a 12-15% annual rate of growth for this end use for aniline will exist at least for the near future.

According to one source (The Society of Dyers and Colourists, 1971), aniline could be used as an intermediate in the production of 174 dyes and dye intermediates, from which many other dyes could be made. Aniline-based dyes include products in the following classes: azo, triphenylmethane, safranine and anthraquinone (Kouris & Northcott, 1963).

Hydroquinone is believed to be the most important photographic chemical derived from aniline, although it is likely that others are also derived from this intermediate. Among the pharmaceuticals and pharmaceutical intermediates derived from aniline are formanilide, acetanilide and sulfanilic acid. Miscellaneous chemicals derived from aniline include sulfonates, toluidines, xylidines, anisidines, phenetidines, nitroanilines, halogenated anilines and N-alkyl- and aryl-derivatives.

(b) Occurrence

Aniline has not been reported to occur as such in nature. Traces of aniline have been detected in cigarette smoke (Neurath et al., 1966; Pailer et al., 1966). It may be present in the waste streams from plants where it is produced or used; and it has been detected in the air 4000 metres from one chemical factory (Burakhovich, 1966).

(c) Analysis

An extensive literature exists concerning the analytical determination of aniline in air, water and in various mixtures including biological materials. The references mentioned below are of particular interest: A method for the detection of aniline in air has been published (HM Factory Inspectorate, 1968). "The Analysis of Air Pollutants" includes a method for aniline (Leithe, 1970), as does "The Determination of Toxic Substances in Air" (Hanson et al., 1965). The photometric determination of aniline in waste water has been described (Zozulya & Mikhailova, 1967). The separation and subsequent detection of a number of amines, including aniline, from a mixture has been carried out using thin-layer and gas chromatography (Sawicki et al., 1966; Uno et al., 1972; Shimomura & Walton, 1968; Kottemann, 1966).

3. Biological Data Relevant to the Evaluation of Carcinogenic Risk to Man

3.1 Carcinogenicity and related studies in animals

In many of the early animal experiments, the purity of the aniline used is unknown, and the experimental techniques employed were often inadequate. Bonser (1947) critically reviewed some of these early studies, paying especial attention to the pathological findings reported; in particular, the positive results reported by Yamazaki & Sato (1937) were considered unreliable. In other studies the period of exposure was often insufficient, possibly due to the toxicity of the doses given. The Working Group agreed with these assessments and considered some of the more adequate studies available.

(a) Oral administration

Rat: Aniline hydrochloride was administered in the drinking-
water of 50 randomly bred rats (sex unspecified) to provide an intake
of 22 mg/day. Half of the rats lived for more than 425 days, the last
rat surviving up to day 750. The total doses administered were
between 14 and 16.5 g/rat. Only the bladder, liver, spleen and
kidneys were examined in all rats. No tumours were observed
(Druckrey, 1950).

(b) Subcutaneous and/or intramuscular injection

Mouse: Of 30 mice, repeated s.c. injections of 1 mg aniline in
lard produced no tumours in 5 mice surviving at 2 years (Shear &
Stewart, 1941). Hartwell & Andervont (1951) gave 8 s.c. injections
of aniline (5 mg in olive oil) to 20 mice, or 13 s.c. injections of
aniline hydrochloride (4 mg in aqueous solution) to 11 mice, and
observed no tumours after 15 and 12 months, respectively. No controls
were reported.

3.2 Other relevant biological data

(a) Animals

All species tested (rabbit, rat, mouse, guinea pig, gerbil,
hamster, cat and dog) oxidize aniline to o- and p-aminophenol, which
are excreted in the urine as various conjugates (Williams, 1959;
Parke, 1960). The ratio of these isomers in the urine differs widely
in the various species. Small amounts of free aniline, phenyl-
sulphamic acid, and aniline-N-glucuronide have been found in the
urines of some species after administration of aniline (Boyland et
al., 1957; Parke, 1960). m-Aminophenol has been detected in trace
quantities in the urine of dogs and rabbits (Parke, 1960).

Aminophenyl- and acetylaminophenylmercapturic acids are also
excreted in rats, and acetylaminophenol and acetylaminophenylmercapturic acid in rabbits (Boyland et al., 1963). Acetanilide has
been detected in the urine of rabbits (0.2%) but not in the urine of
dogs (Williams, 1959). In the rabbit, only traces of metabolites are

excreted in the faeces, and only slight oxidation of the ring carbons occurs to give expired CO_2 (Parke, 1960).

Phenylhydroxylamine (N-Hydroxyaniline) has not been detected in the urine of animals given aniline. However, phenylhydroxylamine and nitrosobenzene appear in the blood of dogs and cats after administration of aniline (Kiese, 1966). The formation of phenylhydroxylamine from aniline appears to be the major cause of the methaemoglobinaemia that follows the administration of this amine (Kiese, 1966). The N-hydroxylation of aniline by hepatic microsomal preparations from several species has been observed in vitro (Kiese, 1966).

(b) Man

The metabolic conversion of aniline to urinary conjugates of p-aminophenol in man has been known since the last century (Williams, 1959). The urinary excretion of these metabolites has been found to give an accurate measure of the absorption of aniline vapour through the skin and respiratory tract in humans (Piotrowski, 1957, 1972; Dutkiewicz, 1961; Dutkiewicz & Piotrowski, 1961). It is probable that other aniline metabolites found in other species appear similarly in the urine of humans exposed to aniline. For example, the methaemoglobinaemia produced in humans by aniline may result from the N-hydroxylation of this amine in vivo. Man is more sensitive than the rat to the methaemoglobinaemia-inducing action of aniline. However, single oral doses of 5 or 15 mg of aniline per person had no effect in 20 adult humans; doses ranging from 25 to 65 mg increased the blood level of methaemoglobin (Jenkins et al., 1972).

(c) Carcinogenicity of metabolites

p-Aminophenol and o-aminophenol hydrochloride each gave negative results when fed to groups of 12-15 rats at concentrations of 0.09-0.2% in the diet for periods of 270 to 341 days (Ekman & Strömbeck, 1949 a, b; Miller & Miller, 1948). o-Aminophenol was ineffective as a carcinogen in mice receiving an implant of this substance in a cholesterol pellet in the lumen of the urinary bladder (Clayson et al., 1958). Phenylhydroxylamine did not induce tumours in groups of 15 to 20 rats given repeated i.p. injections 2 to 3 times per week for 3 months and observed for a period of 1 to 2 years (Miller et al., 1966; Belman et al., 1968). Repeated s.c. injections of 2.5 mg nitrosobenzene twice weekly for 8 weeks did not produce tumours in 20 rats observed up to 8 months after the start of the experiment. No sarcomas occurred at the injection site in 20 control animals, whereas with 2-nitrofluorene local sarcomas occurred in 10 out of 20 rats (Miller et al., 1965).

3.3 Observations in man[1]

The first observation of bladder cancers in industrial workers was made in men using aniline (probably impure) in the manufacture of dyes such as fuchsin (Rehn, 1895). Subsequent early observations of a similar nature and the comparison finding that haemmorrhagic cystitis was common in workers handling aniline lent support to Rehn's view that the bladder tumours were attributable to this amine. These tumours thus became known as "aniline bladder cancers". Later it became evident that 2-naphthylamine and benzidine were far more important in the causation of these bladder cancers than was aniline. Walpole et al. (1954) suggested, on the evidence of their experiments in dogs, that 4-aminodiphenyl (together with 2-naphthylamine and other high-boiling aromatic amines) played a part in the genesis of some of the bladder tumours found in the German dyestuff workers, which had previously been attributed to aniline. In recent decades further English, American, Italian and Russian studies appear to have exonerated aniline as a human carcinogen (Hueper, 1969; Temkin, 1963; Scott, 1962; Vigliani & Barsotti, 1961; Case et al., 1954; Case & Pearson, 1954; Gehrmann et al., 1949).

---

[1] See also preamble, "Carcinogenicity of the aromatic amines in man", of this volume.

4. Comments on Data Reported and Evaluation

4.1 Animal data

There is no adequate data to indicate that aniline is carcinogenic in experimental animals.

4.2 Human data

At the present time, the weight of epidemiological evidence suggests that aniline is not a carcinogen for the human bladder.

5. References

Anon. (1969)   European Chemical News, August 22, 12

Anon. (1971a)   Japan Chemical Annual, p. 52

Anon. (1971b)   European Chemical News, April 23, 14

Anon. (1971c)   Oil, Paint and Drug Reporter, October 18, 9

Anon. (1971d)   Chemical Industry News - India, December, 6

Anon. (1972)   Japan Chemical Week, July 20, 2

Belman, S., Troll, W., Teebor, G. & Mukai, F. (1968)   The carcinogenic and mutagenic properties of N-hydroxy-aminonaphthalenes. Cancer Res., 28, 535

Bonser, G.M. (1947)   Experimental cancer of the bladder. Brit. med. Bull., 4, 973

Boyland, E., Manson, D. & Nery, R. (1963)   Mercapturic acids and metabolites of aniline and 2-naphthylamine. Biochem. J., 86, 263

Boyland, E., Manson, D. & Orr, S.F.D. (1957)   The conversion of arylamines into arylsulphamic acids and arylamine-N-glucosiduronic acids. Biochem. J., 65, 417

Burakhovich, M.C. (1966)   Atmospheric pollution from discharges of the chemical industry. Gig. Sanit., 31, 72

Case, R.A.M., Hosker, M.E., McDonald, D.B. & Pearson, J.T. (1954)   Tumours of the urinary bladder in workmen engaged in the manufacture and use of certain dyestuff intermediates in the British chemical industry. I. The role of aniline, benzidine, alpha-naphthylamine, and beta-naphthylamine. Brit. J. industr. Med., 11, 75

Case, R.A.M. & Pearson, J.T. (1954) Tumours of the urinary bladder in workman engaged in the manufacture and use of certain dyestuff intermediates in the British chemical industry. II. Further consideration of the role of aniline and of the manufacture of auramine and magenta (fuchsine) as possible causative agents. Brit. J. industr. Med., 11, 213

Clayson, D.B., Jull, J.W. & Bonser, G.M. (1958) The testing of ortho-hydroxy-amines and related compounds by bladder implantation and a discussion of their structural requirements for carcinogenic activity. Brit. J. Cancer, 12, 222

Druckrey, H. (1950) Beiträge zur Pharmakologie cancerogener Substanzen. Versuche mit Anilin. Arch. exp. Path. Pharmakol., 210, 137

Dutkiewicz, T. (1961) Aniline vapours absorption in men. Med. Pracy, 12, 1

Dutkiewicz, T. & Piotrowski, J.K. (1961) Experimental investigations on the quantitative estimation of aniline absorption in man. Pure appl. Chem., 3, 319

Ekman, B. & Strömbeck, J.P. (1949a) The effect of feeding aniline on the urinary bladder of rats. Acta path. microbiol. scand., 26, 72

Ekman, B. & Strömbeck, J.P. (1949b) The effect of some split products of 2,3'-azotoluene on the urinary bladder in the rat and their excretion on various diets. Acta path. microbiol. scand., 26, 447

Gehrmann, G.H., Foulger, J.H. & Fleming, A.J. (1949) Occupational carcinoma of the bladder. In: Proceedings of the 9th International Congress of Industrial Medicine, London, 1948, Bristol, Wright, p. 472

Hanson, N.W., Reilly, D.A. & Stagg, H.E. (1965) The Determination of Toxic Substances in Air - A manual of ICI practice, Cambridge, Heffer

Hartwell, J.L. & Andervont, H.B. (1951) In: Shubik, P. & Hartwell, J.L., eds., Survey of compounds which have been tested for carcinogenic activity, Washington, DC, Government Printing Office (US Public Health Publication No. 149), p. 50

HM Factory Inspectorate (1968) Methods for the detection of toxic substances in air - Booklet II Aniline Vapour, London, HMSO

Hueper, W.C. (1969) Occupational and environmental cancers of the urinary system, New Haven, London, Yale University Press, p. 88

Jenkins, F.P., Robinson, J.A., Gellatly, J.B.M. & Salmond, G.W.A. (1972) The no-effect dose of aniline in human subjects and a comparison of aniline toxicity in man and the rat. Fd Cosmet. Toxicol., 10, 671

Kiese, M. (1966) The biochemical production of ferrihemoglobin-forming derivatives from aromatic amines and mechanisms of ferrihemoglobin formation. Pharmacol. Rev., 18, 1091

Kottemann, C.M. (1966) Two-dimensional thin layer chromatographic procedure for the identification of dye intermediates in arylamine oxidation hair dyes. J. Ass. off. analyt. Chem., 49, 954

Kouris, C.S. & Northcott, J. (1963) Aniline and its derivatives. In: Kirk, R.E. & Othmer, D.F., eds., Encyclopedia of Chemical Technology, 2nd ed., New York, John Wiley & Sons, Vol. 2, p. 411

Leithe, W. (1970) The Analysis of Air Pollutants, Ann Arbor, Humphrey Science Publishers

Miller, E.C., Lotlikar, P.D., Pitot, H.C., Fletcher, T.L. & Miller, J.A. (1966) N-hydroxy metabolites of 2-acetylaminophenanthrene and 7-fluoro-2-acetylaminofluorene as proximate carcinogens in the rat. Cancer Res., 26, 2239

Miller, E.C., McKechnie, D., Poirier, M.M. & Miller, J.A. (1965) Inhibition of amino acid incorporation in vitro by metabolites of 2-acetylaminofluorene and by certain nitroso compounds. Proc. Soc. exp. Biol. (N.Y.), 120, 538

Miller, J.A. & Miller, E.C. (1948) The carcinogenicity of certain derivatives of p-dimethylaminoazobenzene in the rat. J. exp. Med., 87, 139

Neurath, G., Dünger, M., Gewe, J., Lüttich, W. & Wichern, H. (1966) Untersuchung der Flüchtigen Basen des Tabakrauches. Beitr. Tabakforsch., 3, 563

Pailer, M., Hübsch, W.J. & Kuhn, H. (1966) Über das Vorkommen aromatischer Amine im Zigarettenrauch (Vorläufige Mitteilung). Mh. Chem., 97, 1448

Parke, D.V. (1960) The metabolism of (^{14}C)- aniline in the rabbit and other animals. Biochem. J., 77, 493

Piotrowski, J.K. (1957) Quantitative estimation of aniline absorption through the skin in man. J. Hyg. Epidem. (Praha), 1, 23

Piotrowski, J.K. (1972) Certain problems of exposure tests for aromatic compounds. Pracov. Lék., 24, 94

Rehn, L. (1895) Blasengeschwülste bei Fuchsin Arbeitern. Arch. klin. Chir., 50, 588

Sawicki, E., Johnson, H. & Kosinksi, K. (1966) Chromatographic separation and spectral analysis of polunuclear aromatic amines and heterocyclic imines. Microchem. J., 10, 72

Scott, T.S. (1962) _Carcinogenic and chronic toxic hazards of aromatic amines_, Amsterdam, New York, Elsevier, p. 49

Shear, M.J. & Stewart, H.L. (1941) In: Shubik, P. & Hartwell, J.L., eds., _Survey of compounds which have been tested for carcinogenic activity_, Washington, DC, Government Printing Office (US Public Health Service Publication No. 149), p. 50

Shimoura, K. & Walton, H.F. (1968) Thin-layer chromatography of amines by ligand exchange. _Separation Science_, _3_, 493

The Society of Dyers and Colourists (1971) _Colour Index_, 3rd ed., _4_, 4699

Temkin, I.S. (1963) _Industrial Bladder Carcinogenesis_, London, Pergamon

US Department of Commerce (March 1967) Business and Defense Services Administration, _Chemicals_, 27

US Department of Commerce (September 1970) Business and Defense Services Administration, _Chemicals_, 25

US Tariff Commission (August 1971) _Imports of Benzenoid Chemicals and Products, 1970_, TC Publication 413

US Tariff Commission (March 1973) _Synthetic Organic Chemicals, United States Production and Sales of Cyclic Intermediates, 1971, Preliminary_

US Tariff Commission (March 14, 1973) _Preliminary Report on US Production of Selected Synthetic Organic Chemicals, Preliminary totals 1972 + January 1973_, SOC Series C/P-73-1

Uno, T., Nakagawa, T. & Toyoda, R. (1972) Reaction gas chromatography - Determination of non-volatile primary amines. _Bunseki Kagaku_, _21_, 993

Vigliani, E.C. & Barsotti, M. (1961) Environmental tumours of the bladder in some Italian dye-stuff factories. _Med. d. Lavoro_, _52_, 241

Walpole, A.L., Williams, M.H.C. & Roberts, D.C. (1952) The carcinogenic action of 4-aminodiphenyl and 3:2'-dimethyl-4-aminodiphenyl. _Brit. J. industr. Med._, _9_, 255

Walpole, A.L., Williams, M.H.C. & Roberts, D.C. (1954) Tumours of the urinary bladder in dogs after ingestion of 4-aminodiphenyl. _Brit. J. industr. Med._, _11_, 105

Williams, R.T. (19590 _Detoxication Mechanisms_, 2nd ed., New York, John Wiley & Sons

Yamazaki, J. & Sato, S. (1937) Experimentelle Erzeugung von Blasengesch-
    wülste durch Anilin und o-Aminoazotoluol bei Kaninchen. <u>Jap. J.
    Dermatol. Urol.</u>, <u>42</u>, 332

Zozulya, A.P. & Mikhailova, L.I. (1967) Photometric determination of
    aniline and m-chloroaniline in waste waters of m-chloroaniline
    production. <u>Khim. Prom.</u>, <u>43</u>, 511

3,3'-DIMETHOXYBENZIDINE[*]

(o-Dianisidine)

1. Chemical and Physical Data

1.1 Synonyms and trade names

Chem. Abstr. No.: 119-90-4

Bianisidine; 4,4'-diamino-3,3'-dimethoxybiphenyl; di-p-amino-di-m-methoxydiphenyl; dianisidine; 3,3'-dimethoxy-4,4'-diaminobiphenyl

1.2 Chemical formula and molecular weight

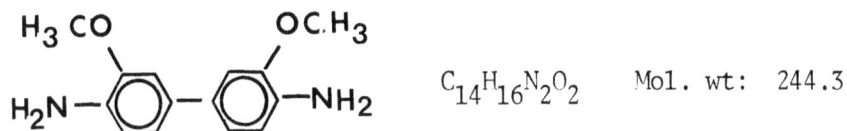

$H_3CO \quad OC.H_3$

$H_2N - \bigcirc - \bigcirc - NH_2$

$C_{14}H_{16}N_2O_2$ Mol. wt: 244.3

1.3 Chemical and physical properties of the pure substance

(a) Description: Colourless crystals which turn violet on standing

(b) Melting-point: 137-138°C

(c) Solubility: Almost insoluble in water, soluble in ethanol, ether, acetone, benzene and chloroform; probably soluble in most organic solvents and lipids

(d) Chemical reactivity: A weak base; has the general characteristics of primary aromatic amines

1.4 Technical products and impurities

It is available commercially as the free base (technical and 99% grades) and as its dihydrochloride.

---

[*] Considered by the Working Group in Lyon, June 1973.

2. Production, Use, Occurrence and Analysis

(a) Production and use[1]

3,3'-Dimethoxybenzidine (o-Dianisidine) has been produced commercially for at least 50 years. It is made commercially by reducing the methyl ether of ortho-nitrophenol (ortho-nitro-anisole) to hydrazoanisole, which is subsequently rearranged by acid to o-dianisidine.

Data on production in the United States of o-dianisidine were last reported for the year 1967, when the total production of five companies amounted to 167 thousand kg (US Tariff Commission, 1968). By 1971, only two US companies were producing o-dianisidine. US imports of o-dianisidine through the principal customs districts were reported to have been 124 thousand kg in 1971 (US Tariff Commission, July 1972).

No data are available on the quantity of o-dianisidine produced in countries other than the United States. In 1967, the Federal Republic of Germany was reported to have one producer; Italy was reported to have one producer in 1969; the United Kingdom was reported to have two producers in 1970; and Japan was reported to have three producers in 1972.

o-Dianisidine (or its dihydrochloride) is used principally as a chemical intermediate for the production of dyes. The next most important application is believed to be as an intermediate in the production of o-dianisidine diisocyanate. It has been reported that o-dianisidine has been used for the detection of the presence of a number of metals, thio-cyanates, and nitrites, and that o-dianisidine itself was formerly used for dyeing acetate rayon (Lurie, 1964).

o-Dianisidine can be used as an intermediate in the production of 89 dyes (The Society of Dyers and Colourists, 1971). Seven of these dyes, for which 1971 production figures have been reported (US Tariff Commission,

---

[1] Data from Chemical Information Services, Stanford Research Institute, USA.

October 1972), are listed in the following table:-

Colour Index No.	Dye	Production (thousands of kg)
24401	Direct Blue 218	479
21160	Pigment Orange 16	153
24410	Direct Blue 1	136
24400	Direct Blue 15	94
24140	Direct Blue 8	64
24411	Direct Blue 76	53
23155	Direct Blue 98	39

In addition, US production of one o-dianisidine-based pigment, Pigment Blue 25 (C.I. 21180), was reported to have been 87 thousand kg in 1971 (US Tariff Commission, August 1972). The o-dianisidine-based dyes and pigments are reportedly useful for dyeing leather, paper, plastics, rubber and textiles (The Society of Dyers and Colourists, 1971).

The phosgenation of the dihydrochloride of o-dianisidine is used in the manufacture of o-dianisidine diisocyanate (also known as 3,3'-dimethoxy-4,4'-diphenylene diisocyanate and 3,3'-dimethoxybenzidine-4,4'-diisocyanate). The amount of o-dianisidine diisocyanate made by the only US producer is not known, but it is estimated that less than 500 thousand kg are produced per year. o-Dianisidine diisocyanate is used in isocyanate-based adhesive systems and as a component of polyurethane elastomers.

(b) Occurrence

o-Dianisidine has not been reported to occur as such in nature. It may be present in the waste streams from plants where it is produced or used. o-Dianisidine is listed as a controlled substance in the UK Carcinogenic Substances Regulations 1967 - Statutory Instrument (1967) No. 879.

(c) Analysis

There are many papers which report methods for the separation and subsequent detection of aromatic amines. A brief overview of this general area can be obtained from the papers on analytical methods cited in the UICC (1970) Monograph. Papers which refer specifically to o-dianisidine usually describe colorimetric detection techniques (Sakai et al., 1960; Glassman & Meigs, 1951). Separation from other amines is normally carried out using thin-layer chromatography (Ghetti et al., 1968).

3. Biological Data Relevant to the Evaluation of Carcinogenic Risk to Man

3.1 Carcinogenicity and related studies in animals

(a) Oral administration

Rat: Pliss (1963, 1965) administered doses of 30 mg o-dianisidine dissolved in sunflower seed oil by stomach tube to rats 3 times per week for 13 months (the original number of rats was not stated). Of 18 surviving rats, 4 developed tumours (2 rats had Zymbal gland tumours, 1 an ovarian tumour and 1 a fibroadenoma of the mammary gland). None of the 50 rats in a control group developed tumours at these sites.

Hadidian et al.(1968) administered this diamine by stomach tube to 30 male and 30 female Fischer rats in doses ranging from 0.1 to 30 mg/animal in a "steroid suspending vehicle" (sodium chloride, sodium carboxymethyl cellulose, polysorbate 80, benzyl alcohol and water) 5 days per week for 52 weeks. Tumours appeared in 293 days, but most were found on autopsy at 18 months. Six rats given the 1 mg doses and 6 rats given the 3 mg doses had a total of 8 tumours; 29 animals receiving the 10 mg doses had a total of 19 tumours; and 6 rats given the 30 mg doses had a total of 5 tumours. Tumours occurred at various sites including the bladder (2 papillomas), the intestine (3 carcinomas), the skin (5 carcinomas) and the Zymbal gland (3 carcinomas). These tumours were not found in 360 control animals given the vehicle alone.

Hamster: Saffiotti et al.(1967) and Sellakumar et al.(1969) fed 0.1% and 1.0% o-dianisidine to groups of 30 male and 30 female Syrian golden hamsters. One urinary bladder tumour occurred in the group receiving the lower level of amine. No primary bladder tumours, but forestomach papillomas, occurred in 37% of animals fed 1% of the amine. Of the control group of hamsters, 2% developed stomach papillomas, but none had urinary bladder tumours.

3.2 Other relevant biological data

(a) Animals

Sciarini & Meigs (1961) found that after administration of a dose of 1 g o-dianisidine to two dogs, 0.4% of free diamine and about 5% of a metabolite with properties similar to those of 3,3'-dihydroxybenzidine were excreted in the urine.

(b) Man

o-Dianisidine has been found in the urine of workers exposed to this compound (Meigs et al., 1951, 1954; Ghetti, 1960).

3.3 Observations in man

No epidemiological data on the occurrence of cancer in workers exposed to o-dianisidine alone appear in the literature. Most of the workers exposed to this diamine have also been exposed to related amines such as benzidine, which has been strongly associated with the occurrence of urinary bladder cancer in man (IARC, 1972).

4. Comments on Data Reported and Evaluation[1]

4.1 Animal data

3,3'-Dimethoxybenzidine (o-Dianisidine) was shown to have a carcinogenic effect in rats following oral administration. The findings obtained in the hamster by the same route suggest a similar effect.

4.2 Human data

No conclusive epidemiological studies have been reported concerning the carcinogenicity of o-dianisidine alone in man.

---

[1] See also the section "Extrapolation from animals to man" in the introduction to this volume.

45

5. References

Ghetti, G. (1960) Escrezione urinaria di alcune ammine aromatiche in lavoratori addetti alla produzione ed all'impiego di benzidina, benzidina sostituite e loro sali. Med. d. Lavoro, 51, 102

Ghetti, G., Bartalini, E., Armeli, G. & Pozzoli, L. (1968) Separazione e dosaggio in vari substrati di amine aromatiche (benzidina, o-tolidina, dianisidina, diclorobenzidina, α-naftilamina, β-naftilamina). Messa a punto di nuovi metodi analitici per l'igiene del lavoro. Lav. umano, 20, 389

Glassman, J.M. & Meigs, J.W. (1951) Benzidine (4,4'-diaminodiphenyl) and substituted benzidines. A microchemical screening technic for estimating levels of industrial exposure from urine and air samples. Arch. industr. Hyg., 4, 519

Hadidian, Z., Fredrickson, T.N., Weisburger, E.K., Weisburger, J.H., Glass, R.M. & Mantel, N. (1968) Tests for chemical carcinogens. Report on the activity of derivatives of aromatic amines, nitrosamines, quinolines, nitroalkanes, amides, epoxides, aziridines and purine antimetabolites. J. nat. Cancer Inst., 41, 985

International Agency for Research on Cancer (1972) IARC Monographs on the Evaluation of Carcinogenic Risk of Chemicals to Man, 1, p.80

Lurie, A.P. (1964) Benzidine and related diaminobiphenyls. In: Kirk, R.E. & Othmer, D.F., eds., Encyclopedia of Chemical Technology, 2nd ed., New York, John Wiley & Sons, Vol. 3, p. 417

Meigs, J.W., Brown, R.M. & Sciarini, L.J. (1951) A study of exposure to benzidine and substituted benzidines in a chemical plant. Arch. industr. Hyg., 4, 533

Meigs, J.W., Sciarini, L.J. & Van Sandt, W.A. (1954) Skin penetration by diamines of the benzidine group. Arch. industr. Hyg., 9, 122

Pliss, G.B. (1963) On some regular relationships between carcinogenicity of aminodiphenyl derivatives and the structure of substance. Acta Un. int. Cancr., 19, 499

Pliss, G.B. (1965) Concerning carcinogenic properties of o-tolidine and dianisidine. Gig. Tr. Prof. Zabol., 9, 18

Saffiotti, U., Cefis, F., Montesano, R. & Sellakumar, A.R. (1967) Induction of bladder cancer in hamsters fed aromatic amines. In: Deichmann, W. & Lampe, K.F., eds., Bladder Cancer. A Symposium, Birmingham, Alabama, Aesculapius, p.129

Sakai, S., Suzuki, K., Mori, H. & Fujino, M. (1960) Color reaction of amines with p-dimethylaminocinnamaldehyde. Japan Analyst, 9, 862

Sciarini, L.J. & Meigs, J.W. (1961) Biotransformation of the benzidines. Arch. environm. Hlth, 2, 584

Sellakumar, A.R., Montesano, R. & Saffiotti, U. (1969) Aromatic amines carcinogenicity in hamsters. Proc. amer. Ass. Cancer Res., 10, 78

The Society of Dyers and Colourists (1971) Colour Index, 3rd ed., 4, 2221, 2223, 2226, 2248, 2256, 2299, 3292, 3293, 3348, 3349, 4742

UICC (1970) The quantification of environmental carcinogens (UICC Technical Report Series, Vol. 4)

US Tariff Commission (1968) Synthetic Organic Chemicals, United States Production and Sales, 1967, TC Publication 295

US Tariff Commission (July 1972) Imports of Benzenoid Chemicals and Products, 1971, TC Publication 466

US Tariff Commission (August 1972) Synthetic Organic Chemicals, United States Production and Sales of Organic Pigments, 1971 Preliminary

US Tariff Commission (October 1972) Synthetic Organic Chemicals, United States Production and Sales of Dyes, 1971 Preliminary

3,3'-DICHLOROBENZIDINE*

1. Chemical and Physical Data

1.1 Synonyms and trade names

Chem. Abstr. No.: 91-94-1

C.I. 23060 (Colour Index); DCB; 4,4'-diamino-3,3'-dichlorobiphenyl; o,o'-dichlorobenzidine; dichlorobenzidine; dichlorobenzidine base; 3,3'-dichloro-4,4'-biphenyldiamine; 3,3'-dichlorobiphenyl-4,4'-diamine

1.2 Chemical formula and molecular weight

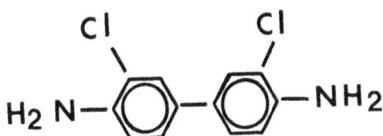

$C_{12}H_{10}Cl_2N_2$ Mol. wt: 253.1

1.3 Chemical and physical properties of the pure substance

(a) Description: Colourless crystals

(b) Melting-point: 132-133°C

(c) Solubility: Almost insoluble in cold water; readily soluble in ethanol, benzene, glacial acetic acid; slightly soluble in dilute hydrochloric acid

(d) Chemical reactivity: A weak base, has the general characteristics of primary aromatic amines

1.4 Technical products and impurities

Dichlorobenzidine (DCB) is believed to be available commercially as the free base, but no information could be obtained as to its purity. The

---

* Considered by the Working Group in Lyon, June 1973.

dihydrochloride salt is available with a minimum content of dichlorobenzidine base of 60% (this corresponds to only 77.3% of the theoretical
base content of the pure dihydrochloride) and 0.2%, maximum, of material
insoluble in hydrochloric acid[1].

2. Production, Use, Occurrence and Analysis

(a) Production and use[1]

DCB has been produced commercially for at least 35 years (US Tariff
Commission, 1938). It can be produced by the following two methods (the
method used for commercial production could not be established): (i) reduction of ortho-nitrochlorobenzene to the hydrazo compound and rearrangement of this to DCB using mineral acid, and (ii) chlorination of diacetyl-
benzidine with hypochlorite followed by hydrolysis of the diamide to the
diamine using hydrochloric acid.

Combined production in the United States (by three companies) of the
DCB base and its salts was reported to have been 1.6 million kg in 1971
(US Tariff Commission, March 1973). In 1971 US imports through the
principal customs districts were reported to have been 658 thousand kg,
mainly from Japan and the Federal Republic of Germany (US Tariff Commission, July 1972). In 1969, Italy was reported to have one producer of
DCB, and Japan was reported to have three producers in 1972.

DCB is used principally as a chemical intermediate for the production
of dyes and pigments. In this application the salts (e.g., the hydrochloride or the sulphate) may be preferred to the free base. The next
most important application is believed to be as a curing agent for isocyanate-containing polymers. It has been reported that DCB has also found
use in a colour test for the presence of gold (Lurie, 1964).

DCB can be used as an intermediate in the production of thirteen dyes
or pigments (The Society of Dyers and Colourists, 1971). The only six
products for which the 1971 US production figures have been reported (US
Tariff Commission, August 1972) are listed in the following table:-

---

[1] Data from Chemical Information Services, Stanford Research
Institute, USA.

50

Colour Index No.	Dye or Pigment	Production (thousands of kg)
21090	Pigment Yellow 12	2,547
21095	Pigment Yellow 14	1,034
21105	Pigment Yellow 17	260
21110	Pigment Orange 13	65
21115	Pigment Orange 34	55
21120	Pigment Red 38	60

Production of Pigment Yellow 13 (C.I. 21100) was reported to have been 199 thousand kg in 1970 (US Tariff Commission, 1972). The DCB-based pigments are reportedly useful for colouring most common plastic resins, rubber, printing inks, metal finishes, and in textile and wallpaper printing (The Society of Dyers and Colourists, 1971).

DCB is used alone and in blends with 4,4'-methylenebis(2-chloro-aniline) as a curing agent for liquid-castable polyurethane elastomers. Since it is used in smaller quantities than is 4,4'-methylenebis(2-chloro-aniline), US consumption of DCB as a curing agent is believed to be no more than a few hundred thousand kg per year.

(b) Occurrence

DCB has not been reported to occur as such in nature. It may be present in the waste streams from plants where it is produced or used. Since less than stoichiometric amounts are usually used in the production of cured polyurethane elastomers with DCB, unreacted diamine is not normally present. However, the curing agents are often melted before mixing into the elastomer formulations, so DCB could possibly be volati-lized and find its way into the waste gases and water.

DCB is listed as a controlled substance in the UK Carcinogenic Substances Regulations 1967 - Statutory Instrument (1967) No. 879. DCB is also listed with 13 other compounds in the US Federal Register (US Government, 1973) as being subject to an Emergency Temporary Standard on certain carcinogens under an order made by the Occupational Safety and Health Administration, Department of Labor on 26 April, 1973.

(c) Analysis

Ghetti et al. (1968) describe methods for the analysis of DCB in a factory environment, including a method capable of detecting 5 µg/100 ml of urine. Similar findings are reported by Akiyama (1970).

The reaction with p-dimethylamino-benzaldehyde is also used to detect DCB (Sakai et al., 1960). In addition, general methods for detecting aromatic amines can be employed (UICC, 1970).

3. Biological Data Relevant to the Evaluation of Carcinogenic Risk to Man

3.1 Carcinogenicity and related studies in animals

(a) Oral administration

Rat: Pliss (1959) observed a high incidence of adenomas and carcinomas of the Zymbal gland and other organs including 2 tumours of the bladder in 12 out of 50 rats fed 10-20 mg DCB in the diet 6 times per week, for 12 months (total dose 4.5 g/rat).

Stula et al. (1971, 1973) reported on a test lasting 16 months which revealed that when 1,000 ppm DCB were fed to 50 male and 50 female ChR CD rats in a standard diet (23% protein), malignant mammary, skin and acoustic duct tumours developed in both sexes, and an excess of haemopoietic tumours arose in the males compared to their incidence in an equal number of controls.

Hamster: In lifetime studies in this species, dietary levels of 0.1% DCB in the diet did not induce tumours in 30 male and 30 female Syrian golden hamsters, when compared with a similar number of untreated animals. However, in later studies in similar groups of animals 0.3% DCB in the diet produced 4 transitional cell carcinomas of the bladder and some liver cell tumours; these tumours were not found in the control animals (Saffiotti et al., 1967; Sellakumar et al., 1969).

(b) Subcutaneous and/or intramuscular injection

Rat: Pliss (1963) injected a group of rats s.c. with 15-60 mg

DCB/rat in sunflower seed oil or glycerol and water at unspecified intervals for 10 to 13 months. The total percentage of tumour-bearing animals was 74%. Skin, sebaceous and mammary gland tumours were observed most frequently, and there were also intestinal, urinary bladder and bone tumours. Among 25 control rats injected with the vehicle alone, only one tumour was observed, namely, a sarcoma connected with the wall of a parasitic cyst.

3.2 Other relevant biological data

(a) Animals

Kellner et al. (1973) compared the metabolism of benzidine and DCB in rats, dogs and monkeys after intravenous injection. Benzidine was excreted much more rapidly than was the dichloro-derivative.

3.3 Observations in man

No case reports in which DCB has induced cancer in man are known. However, DCB may have contributed to cases of bladder cancer attributed to benzidine, as both substances may be prepared in the same plant.

4. Comments on Data Reported and Evaluation[1]

4.1 Animal data

3,3'-Dichlorobenzidine (DCB) is carcinogenic in the rat following oral and subcutaneous administration and in the hamster after oral administration.

4.2 Human data

No epidemiological data are available, but as 3,3'-dichlorobenzidine and benzidine may be made in the same plant, the possibility cannot be excluded that dichlorobenzidine has contributed to the incidence of bladder cancer attributed to benzidine.

---

[1] See also the section "Extrapolation from animals to man" in the introduction to this volume.

5. References

Akiyama, T. (1970) The investigation on the manufacturing plant of organic pigment. Jikei med. J., 17, 1

Ghetti, G., Bartalini, E., Armeli, G. & Pozzoli, L. (1968) Separazione e dosaggio in vari substrati di amine aromatiche (benzidina, o-tolidina, dianisidina, dichlorobenzidina, α-naftilamina, β-naftilamina). Messa a punto di nuovi metodi analitici per l'igiene del lavoro. Lav. umano, 20, 389

Kellner, H.M., Christ, O.E. & Lötzsch, K. (1973) Animal studies on the kinetics of benzidine and 3,3'-dichlorobenzidine. Arch. Toxikol. (in press)

Lurie, A.P. (1964) Benzidine and related diaminobiphenyls. In: Kirk, R.E. & Othmer, D.F., eds., Encyclopedia of Chemical Technology, 2nd ed., New York, John Wiley & Sons, Vol. 3, p. 146

Pliss, G.B. (1959) The blastomogenic action of dichlorobenzidine. Vop. Onkol., 5, 524

Pliss, G.B. (1963) On some regular relationships between carcinogenicity of aminodiphenyl derivatives and the structure of substance. Acta Un. int. Cancr, 19, 499

Saffiotti, U., Cefis, F., Montesano, R. & Sellakumar, A.R. (1967) Induction of bladder cancer in hamsters fed aromatic amines. In: Deichmann, W. & Lampe, K.F., eds., Bladder Cancer. A Symposium, Birmingham, Alabama, Aesculapius, p. 129

Sakai, S., Suzuki, K., Mori, H. & Fujino, M. (1960) Colour reaction of amines with p-dimethylamino-benzaldehyde. Japan Analyst, 9, 862

Sellakumar, A.R., Montesano, R. & Saffiotti, U. (1969) Aromatic amines carcinogenicity in hamsters. Proc. amer. Assoc. Cancer Res., 10, 78

The Society of Dyers and Colourists (1971) Colour Index, 3rd ed., 4, 3272-3275, 3290-3291, 3294-3295, 3304-3305, 4742

Stula, E.F., Sherman, H. & Zapp, J.A., Jr (1971) Experimental neoplasia in ChR-CD rats with the oral administration of 3,3'-dichlorobenzidine, 4,4'-methylenebis(2-chloroaniline) and 4,4'-methylenebis(2-methylaniline). Toxicol. appl. Pharmacol., 19, 380

Stula, E.F., Sherman, H., Zapp, J.A., Jr & Clayton, J.W., Jr (1973) Experimental neoplasia in rats from oral administration of 3,3'-dichlorobenzidine, 4,4'-methylene-bis(2-chloroaniline) and 4,4'-methylene-bis(2-methylaniline). Toxicol. appl. Pharmacol. (in press)

UICC (1970) The quantification of environmental carcinogens (UICC
     Technical Report Series, Vol. 4)

US Government (1973) Occupational safety and health standards. US
     Federal Register, 38, No. 85, 10929

US Tariff Commission (1938) Dyes and Other Synthetic Organic Chemicals in
     the United States, 1937, Report No. 132, Second Series

US Tariff Commission (July 1972) Imports of Benzenoid Chemicals and
     Products, 1971, TC Publication 466

US Tariff Commission (1972) Synthetic Organic Chemicals, United States
     Production and Sales, 1970, TC Publication 479

US Tariff Commission (August 1972) Synthetic Organic Chemicals, United
     States Production and Sales of Organic Pigments, 1971 Preliminary

US Tariff Commission (March 1973) Synthetic Organic Chemicals, United
     States Production and Sales of Cyclic Intermediates, 1971 Preliminary

MAGENTA[*]

1. Chemical and Physical Data

Magenta is a mixture of three closely related 4,4',4''-triaminotriaryl-methane dyes in the form of their monohydrochloride salts.

1.1 Synonyms and trade names[1]

Chem. Abstr. No.: MX8053096

Basic fuchsine; basic magenta; C.I.42510 (Colour Index); C.I. Basic Violet 14 (Colour Index); fuchsine

1.2 Chemical formula and molecular weight

(a) 2-Methyl-4,4'-((4-imino-2,5-cyclohexadien-1-ylidene)methylene) dianiline monohydrochloride, also known as fuchsine; magenta I; rosaniline[2]

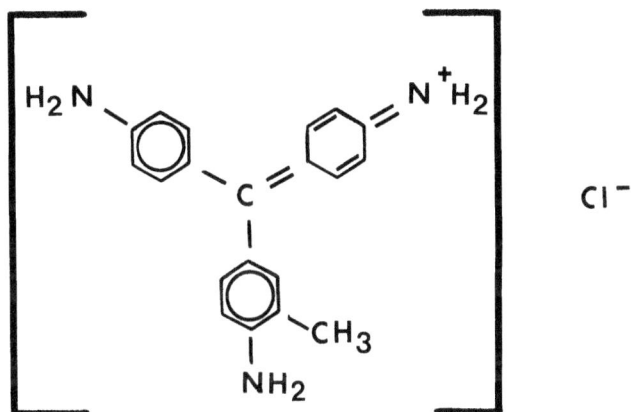

$C_{20}H_{18}N_3$. HCl[.]

Mol. wt: 337.8

---

[*] Considered by the Working Group in Lyon, June 1973.

[1] Caution should be used in referring to these materials, since the term "magenta" appears to be used quite loosely in the literature.

[2] Alternative names are sometimes used for these chemicals, in that they can be considered to be derivatives of 4-toluidine or of 2,4-xylidene.

(b) 4,4'-((4-imino-2,5-cyclohexadien-1-ylidene)methylene)dianiline monohydrochloride, also known as parafuchsine; para-magenta; para-rosaniline[1]

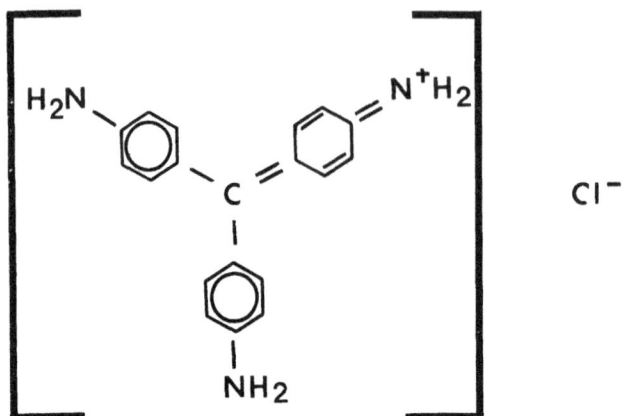

$C_{19}H_{16}N_3$ · HCl

Mol. wt: 323.8

(c) 4,4'-((4-imino-2,5-cyclohexadien-1-ylidene)methylene)di-o-toluidine monohydrochloride, also known as magenta II[1]

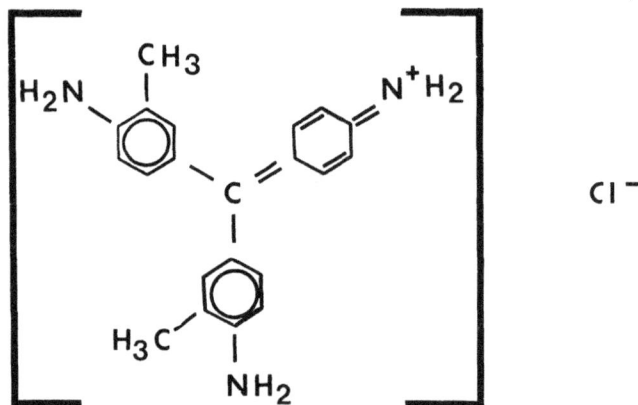

$C_{21}H_{20}N_3$ · HCl

Mol. wt: 351.9

---

[1] Alternative names are sometimes used for these chemicals, in that they can be considered to be derivatives of 4-toluidine or of 2,4-xylidine.

1.3 Chemical and physical properties of the commercial product

(a) <u>Description</u>:   Brownish red crystals

(b) <u>Melting-point</u>:   Decomposes at about $186^{o}C$

(c) <u>Solubility</u>:   Slightly soluble in water, soluble in alcohols and acids; almost insoluble in ether

(d) <u>Chemical reactivity</u>:   Easily reduced to colourless leuco-bases

1.4 Technical products and impurities

Magenta, as a commercial material, is a mixture of the three closely related chemicals specified above. It contains mostly compound (a), some of compound (b) and a small amount of compound (c) (Witterholt, 1969).

Magenta is also available in the United States as Basic Fuchsin, N.F. grade. The specifications for this material set a maximum limit of 8 parts per million of arsenic and 30 parts per million of lead (American Pharmaceutical Association, 1970).

2. Production, Use, Occurrence and Analysis

One review concerning magenta and related triphenylmethane dyes has been published (Witterholt, 1969).

(a) <u>Production and use</u>[1]

Magenta was discovered in 1856 and was being produced for sale in a factory in England before 1874 (Bannister & Olin, 1965). It has been produced commercially in the United States for at least 50 years (US Tariff Commission, 1922). The preferred commercial production method is the oxidation of a mixture of aniline, ortho- and para-toluidine and their hydrochlorides by nitrobenzene or by ortho-nitrotoluene, or a mixture of the two, in the presence of ferrous chloride, ferrous oxide and zinc chloride. In the past, magenta has also been made by heating a mixture of

---

[1] Data from Chemical Information Services, Stanford Research Institute, USA.

aniline and ortho- and para-toluidine with arsenic acid.

Separate US production data for magenta were last reported for 1964 when the combined production of five US producers was reported to have been 53 thousand kg (US Tariff Commission, 1965). Total US production of a group of "All other basic violet dyes", which included magenta (made by two producers) and nine other dyes, was reported to have been 700 thousand kg in 1971 (US Tariff Commission, October 1972).

US imports of magenta through the principal customs districts decreased from the 11 thousand kg reported for 1969 (US Tariff Commission, July 1970) to 3 thousand kg in 1971 (US Tariff Commission, July 1972).

No data are available on the quantity of magenta produced in countries other than the US. In 1971, it was reported that magenta was made by six companies in the United Kingdom, and by one company each in Japan, Switzerland and The Netherlands (The Society of Dyers and Colourists, 1971).

Magenta is believed to be used mainly as a dye and, to a lesser extent, as a dye intermediate. Its major commercial applications are probably in the colouring of textile fibres, fabrics and paper products. The most significant use of magenta when employed alone is believed to be in the dyeing of cotton and wool; however, it has been reported to be a constituent of an important, three-component black dye which, because of its light-fastness, is widely used on modified acrylic fibres (Witterholt, 1969). Magenta has also been reported as being used in the dyeing of bast (jute) fibres, china clay products and leather; in printing inks; and as a filter dye in photography (The Society of Dyers and Colourists, 1971).

Although the quantity consumed is probably quite small, magenta is widely used in medicine as a microbiological stain. One source has described the basic fuchsines (magenta and related dyes) as among the most powerful nuclear dyes and has listed 26 biological stains in which they are used (Witterholt, 1969). Magenta has been reported to be in use in meat-marking colours in New Zealand (Dacre, 1971); and it has additional uses as a laboratory reagent, e.g., in thin-layer chromatography.

The grade of magenta used in medicine, Basic Fuchsin, N.F., is approved for use in Carbol-Fuchsin Solution, N.F., an antifungal agent containing phenol and resorcinol. No evidence is available to indicate that this solution is presently being used in the US, but it is probably used elsewhere.

Magenta is used in a variety of ways as a dye intermediate. Conversion of the monohydrochlorides to the free bases results in C.I. Solvent Red 41 (C.I. 42510:1), which is reportedly used in polishes, carbon paper and ballpoint-pen inks. Conversion to the phosphotungstomolybdic acid salts results in C.I. Pigment Violet 4 (C.I. 42510:2), which has been reported to find use as a pigment in printing inks (The Society of Dyers and Colourists, 1971). Production of C.I. Solvent Red 41 was reported by only one US company in 1971, and no US companies reported production of C.I. Pigment Violet 4 (US Tariff Commission, October 1972).

(b) Occurrence

Magenta has not been reported to occur as such in nature. It may be present in the waste streams from plants where it is made or used.

Magenta is listed as a controlled substance in the UK Carcinogenic Substances Regulations 1967 - Statutory Instrument (1967) No. 879.

(c) Analysis

Spectrophotometric methods are described by Emery & Stotz (1953), but absorption maxima differ slightly for the individual components of magenta. Its determination in pharmaceutical preparations has been described by Sukhlitskaya & Belevich (1963). Guidelines for the analysis of aromatic amines are also available (UICC, 1970).

3. Biological Data Relevant to the Evaluation of Carcinogenic Risk to Man

3.1 Carcinogenicity and related studies in animals

(a) Oral administration

Mouse: Bonser et al.(1956) gave 12 mg magenta in arachis oil per week to 60 stock mice by gastric instillation for 52 weeks (total dose, 624 mg). Dye was found to have stained the tissues at autopsy.

61

Seven of the 60 mice were examined before week 50, compared with 2 of 60 controls; and 14 mice in the test group were examined between weeks 50 and 90, compared with 16 mice in the control group. Four lymphomas and 1 hepatoma were found in the 21 test animals, and 5 lymphomas were found in 18 controls. A further 8 mice were examined for intestinal tumours between weeks 50 and 90, but none were found. Many mice died from ectromelia throughout the experiment.

(b) Subcutaneous and/or intramuscular injection

Rat: Druckrey et al. (1956) injected 10 mg 4,4',4''-triaminotriphenylmethane hydrochloride (para-magenta) as a 1% aqueous solution s.c. at weekly intervals into 20 BD III rats. The first local sarcoma appeared at 300 days, after a total dose of 370 mg of the dye. A total of 7 sarcomas was found in 12 rats surviving after the appearance of the first tumour, compared to a spontaneous incidence of sarcomas in these rats of less than 0.5%.

3.2 Other relevant biological data

No information is available to the Working Group.

3.3 Observations in man

(a) Case reports: Rehn (1895) was the first to report the appearance of tumours associated with the manufacture of this substance. Other reports are reviewed by Hueper (1942).

(b) Epidemiology: Case & Pearson (1954), in a survey of the British chemical industry, showed that there was a highly significant risk of contracting bladder cancer among men engaged in manufacturing, but not in purifying or using, magenta. Of 85 men known to have been engaged in manufacturing magenta, but not exposed to 1- or 2-naphthylamine or benzidine, there were 5 cases of bladder cancer. Three of these were recorded as dying of bladder cancer, whereas only 0.13 cases would have been expected (relative risk = 23.0; $P < 0.005$).

It is known, however, that the modern process for manufacturing magenta involves the replacement of aniline by o-toluidine, and there is speculation that o-toluidine may be implicated in the aetiology of the magenta tumours.

4. Comments on Data Reported and Evaluation

4.1 Animal data

The only evidence for the carcinogenicity of magenta is the induction of local sarcomas in rats following subcutaneous administration of para-magenta, one of the components of commercial magenta. This positive result may have been due to the physical rather than to the chemical properties of this substance. Oral administration of commercial magenta to mice produced negative effects in a single study, but this finding was based upon an insufficient number of surviving animals.

4.2 Human data

One epidemiological study appears to establish the carcinogenic risk to workers involved in the manufacture of magenta. On present evidence, it is not possible to indicate whether the industrial bladder cancer found in magenta workers is attributable to exposure to magenta itself, or to one or more of its associated intermediates and impurities.

5. References

American Pharmaceutical Association (1970) National Formulary XIII, Monographs 139 and 322

Bannister, D.W. & Olin, A.D. (1965) Dyes and dye intermediates. In: Kirk, R.E. & Othmer, D.F., eds., Encyclopedia of Chemical Technology, 2nd ed., New York, John Wiley & Sons, Vol. 7, p. 463

Bonser, G.M., Clayson, D.B. & Jull, J.W. (1956) The induction of tumours of the subcutaneous tissues, liver and intestine in the mouse by certain dyestuffs and their intermediates. Brit. J. Cancer, 10, 653

Case, R.A.M. & Pearson, J.T. (1954) Tumours of the urinary bladder in workmen engaged in the manufacture and use of certain dyestuff intermediates in the British chemical industry. Part II. Further consideration of the role of aniline and of the manufacture of auramine and magenta (fuchsine) as possible causative agents. Brit. J. industr. Med., 11, 213

Dacre, J.C. (1971) Carcinogenic compounds on edible meat in New Zealand. N.Z. med. J., 73, 74

Druckrey, H., Nieper, H.A. & Lo, H.W. (1956) Carcinogene Wirkung von Parafuchsin im Injektionsversuch an Ratten. Naturwissenschaften, 43, 543

Emery, A.J. & Stotz, E. (1953) Spectrophotometric characteristics and assay of biological stains. IV. The phenylmethane dyes. Stain Technol., 28, 235

Hueper, W.C. (1942) Occupational Tumours and Allied Diseases, Springfield, Illinois, Thomas

Rehn, L. (1895) Blasengeschwülste bei Fuchsinarbeitern. Arch. klin. Chir., 50, 588

The Society of Dyers and Colourists (1971) Colour Index, 3rd ed., 4, 1652, 3334-3335, 3596, 4389, 5079-5080, 5573-5574, vii-xi

Sukhlitskaya, Y.M. & Belevich, N.A. (1963) Fuchsine determination in drugs. Aptechn. Delo., 12, 75

UICC (1970) The quantification of environmental carcinogens (UICC Technical Report Series, Vol. 4)

US Tariff Commission (1922) Census of Dyes and other Synthetic Organic Chemicals, 1921, Tariff Information Series No. 26

US Tariff Commission (1965) Synthetic Organic Chemicals, United States Production and Sales, 1964, TC Publication 167

US Tariff Commission (July 1970) Imports of Benzenoid Chemicals and Products, 1969, TC Publication 328

US Tariff Commission (July 1972) Imports of Benzenoid Chemicals and Products, 1971, TC Publication 496

US Tariff Commission (October 1972) Synthetic Organic Chemicals, United States Production and Sales of Dyes, 1971 Preliminary

Witterholt, V.G. (1969) Triphenylmethane and related dyes. In: Kirk, R.E. & Othmer, D.F., eds., Encyclopedia of Chemical Technology, 2nd ed., New York, John Wiley & Sons, Vol. 20, p. 672

4,4'-METHYLENE BIS(2-CHLOROANILINE)*

1. Chemical and Physical Data

1.1 Synonyms and trade names

Chem. Abstr. No.: 101-14-4

Bis amine; di-(4-amino-3-chlorophenyl)methane; 4,4'-diamino-3-3'-dichlorodiphenylmethane; 3,3'-dichloro-4,4'-diaminodiphenylmethane; mboca; methylenebis(ortho-chloroaniline); p,p'-methylenebis(ortho-chloroaniline)

Curalin M; Curene 442; Cyanaset; MOCA; DACPM

1.2 Chemical formula and molecular weight

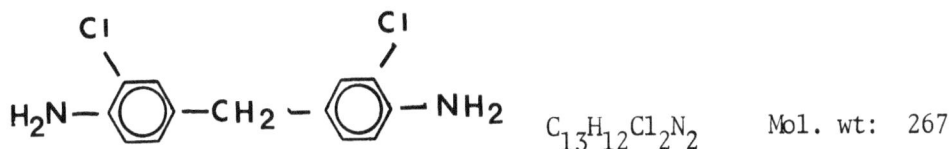

$C_{13}H_{12}Cl_2N_2$ Mol. wt: 267

1.3 Chemical and physical properties of the pure substance

(a) Description: Colourless crystals

(b) Melting-point: $110^{o}C$

(c) Solubility: Almost insoluble in water; soluble in alcohol and ether and probably in most organic solvents and lipids

(d) Chemical reactivity: A weak base; has the general characteristics of primary aromatic amines

1.4 Technical products and impurities

4,4'-Methylenebis(2-chloroaniline), frequently referred to as DACPM, is available commercially in the United States in the form of pellets, but

---

* Considered by the Working Group in Lyon, June 1973.

no information is available on their purity. Some 2-chloro-4-methyl-aniline might be present as a result of the process of manufacture.

2. Production, Use, Occurrence and Analysis

(a) Production and use[1]

Production of DACPM was first reported to the US Tariff Commission in 1956 (US Tariff Commission, 1957). Commercial production is believed to be based on the reaction of formaldehyde with ortho-chloroaniline. It is estimated that in 1970 US production was in the order of 1.5-2.5 million kg and that 1972 production was approximately 3.5 million kg. There have been indications in the literature that two companies produce this chemical. In 1970 US imports through the principal customs districts amounted to approximately 1000 kg (US Tariff Commission, August, 1971).

No data are available on the quantity of DACPM produced in countries other than the US. In 1970, the United Kingdom was reported to have one producer. Japan was reported to have three producers in 1972.

Virtually all (perhaps 99%) of the DACPM consumed in the US is believed to be used as a curing agent for isocyanate-containing polymers; and it is reported to be the most widely used agent for curing liquid-castable polyurethane elastomers suitable for molded mechanical articles and for potting and encapsulating purposes (Gianatasio, 1969). It is frequently formulated with other aromatic diamines (e.g., 3,3'-dichloro-benzidine or 4,4'-methylenedianiline) to prepare curing agents sold under trade names.

Very small quantities (perhaps 1% of the total consumed) of DACPM are believed to be used as a curing agent for epoxy and epoxy-urethane resin blends.

(b) Occurrence

DACPM has not been reported to occur as such in nature. Since less

---

[1] Data from Chemical Information Services, Stanford Research Institute, USA.

than stoichiometric amounts are usually used in the production of poly-urethane elastomers cured with DACPM, unreacted diamine is not normally present. The curing agents are melted before mixing into the elastomer formulations, thus DACPM could possibly be volatilized and find its way into the waste gases and water from plants where it is being used.

DACPM is listed as a controlled substance in the UK Carcinogenic Substances Regulations 1967 - Statutory Instrument (1967) No. 879.

DACPM is also listed with 13 other compounds in the US Federal Register (US Government, 1973) as being subject to an Emergency Temporary Standard on certain carcinogens under an order made by the Occupational Safety and Health Administration, Department of Labor, on 26 April, 1973.

(c)   Analysis

Guidelines for the analysis of aromatic amines are available (UICC, 1970).

Techniques for quantifying the amounts of DACPM which may be present in working atmospheres and in the urine of workers handling this compound have recently been developed, using gas, paper and thin-layer chromato-graphy. The maximum limit of sensitivity by thin-layer chromatography is 0.04 mg/l (Linch et al., 1971). A colorimetric method for the determina-tion of DACPM in air which is also contaminated with aromatic isocyanates has been described (Meddle & Wood, 1970).

3.   Biological Data Relevant to the Evaluation of Carcinogenic Risk to Man

3.1   Carcinogenicity and related studies in animals

(a)   Oral administration

Mouse: Russfield et al. (1973) tested DACPM in mice at levels of 0.2% and 0.1% in the diet. Groups of 25 males and 25 females were used at each dose level and as controls. Vascular tumours were the most striking finding, being present in 40% of males at the higher dose level, in 23% at the lower level and being absent in the controls. In female mice these vascular tumours (43%) were found only at the higher dose level, while hepatomas occurred in 50% of females at the higher dose, in 43% at the lower dose and not at all in controls.

Rat: One group of 25 male and 25 female rats was maintained throughout their lifespan on a low protein diet containing 0.1% DACPM. The total dose administered was 27 g/kg bw. The average survival time was 565 days in male rats and 535 days in female rats. Of the male rats, 23 died with tumours (22 with multifocal hepatomas); of the females, 20 died with tumours (18 with multifocal hepatomas). Primary lung tumours (mainly carcinomas) were observed in 13 animals, 10 of which also had hepatomas. Controls on a similar diet and sur- viving up to 730 days produced only 2 mammary adenomas (Grundmann & Steinhoff, 1970).

Stula et al. (1971, 1973) maintained 50 male and 50 female ChR- CD rats on a standard diet (23% protein) containing 1000 ppm DACPM. The average number of test days was 560 for males and 548 for the females; for controls the corresponding values were 564 and 628 days. Primary lung adenomas were seen at one year, and later adenocarcino- mas occurred, occasionally with distant metastases. A few instances of squamous cell neoplasias and, rarely, pleural mesotheliomas appeared. The overall yield was 83 lung tumours in 100 animals, 5 of which had pleural mesotheliomas. Liver tumours occurred in both sexes; and in both groups miscellaneous tumours were also more fre- quently found. In rats maintained for up to 15 months on a low protein (7%) diet, 35 out of 50 developed lung tumours, and 1 developed, in addition, a pleural mesothelioma. Some tumours were more frequent in test males than in controls, while in females the incidence of mammary tumours was significantly higher in treated than in control rats.

Russfield et al. (1973) administered 0.1% and 0.05% DACPM (97% pure) to male ChR-CD-1 rats in the diet; 25 male rats were treated at each dose level. The only dose related response was the occur- rence of 4 hepatomas in 4/19 rats on the higher dose level surviving at the appearance of the first tumour, and 1 hepatoma in 1/22 rats at the lower level, with none in the controls.

(b) Subcutaneous and/or intramuscular injection

Rat: DACPM (94% pure) in physiological saline was given weekly by repeated s.c. injections to 17 male and 17 female rats over a period of 620 days (total dose, 25 g/kg bw). The rats received a laboratory diet with normal protein content. Nine rats developed liver cell carcinomas (8 multifocal), and 7 rats had primary lung carcinomas (3 multifocal). Only 1 benign tumour was found in the s.c. tissues. A total of 13 tumours occurred in 50 control rats surviving up to 1040 days, but no malignant tumours of the liver or lungs were seen (Steinhoff & Grundmann, 1971).

3.2 Other relevant biological data

(a) Man

DACPM has been identified in the urine of workers exposed to this compound (Linch et al., 1971).

3.3 Observations in man

(a) Epidemiology

In a recent study of 31 men whose exposure to DACPM ranged from 6 months to 16 years, Linch et al. (1971) found no cytological evidence of a bladder cancer hazard. In addition, 178 other DACPM workers with unstated duration of exposure were also examined cytologically more than 10 years after exposure had ceased, and no cases of bladder cancer were found.

4. Comments on Data Reported and Evaluation[1]

4.1 Animal data

4,4'-Methylene bis(2-chloroaniline) is carcinogenic in the mouse and

---

[1] See also the section "Extrapolation from animals to man" in the introduction to this volume.

rat after oral administration and produces distant tumours in the rat after subcutaneous administration.

4.2 Human data

There are no conclusive epidemiological studies on which an evaluation of the carcinogenic risk can be based.

5. References

Gianatasio, P.A. (1969) Polyurethane polymers. Part I; chemistry and characteristics. Rubber Age, July, 51

Grundmann, E. & Steinhoff, D. (1970) Leber und Lungentumoren nach 3,3'-Dichlor-4,4'-diaminodiphenylmethan bei Ratten. Z. Krebsforsch., 74, 28

Linch, A.L., O'Connor, G.B., Barnes, J.R., Killian, A.S., Jr & Neeld, W.E., Jr (1971) Methylene-bis-ortho-chloroaniline (MOCA[R]): Evaluation of hazards and exposure control. Amer. industr. Hyg. Ass. J., 32, 802

Meddle, D.W. & Wood, R. (1970) A method for the determination of aromatic isocyanates in air in the presence of primary aromatic amines. Analyst, 95, 402

Russfield, A.B., Homburger, F., Boger, E., Weisburger, E.K. & Weisburger, J.H. (1973) The carcinogenic effect of 4,4'-methylene-bis-(2-chloro-aniline) in mice and rats. Toxicol. appl. Pharmacol. (in press)

Steinhoff, D. & Grundmann, E. (1971) Zur cancerogen Wirkung von 3,3'-Dichlor-4,4'-diaminodiphenylmethan bei Ratten. Naturwissenschaften, 58, 578

Stula, E.F., Sherman, H. & Zapp, J.A., Jr (1971) Experimental neoplasia in ChR-CD rats with the oral administration of 3,3'-dichlorobenzidine, 4,4'-methylenebis(2-chloroaniline), and 4,4'-methylene-bis(2-methyl-aniline). Toxicol. appl. Pharmacol., 19, 380

Stula, E.F., Sherman, H., Zapp, J.A., Jr & Clayton, J.W., Jr (1973) Experimental neoplasia in rats from oral administration of 3,3'-dichloro-benzidine, 4,4'-methylene-bis(2-chloroaniline) and 4,4'-methylene-bis-(2-methylaniline). Toxicol. appl. Pharmacol. (in press)

UICC (1970) The quantification of environmental carcinogens (UICC Technical Report Series, Vol. 4)

US Government (1973) Occupational safety and health standards. US Federal Register, 38, No. 85, 10929

US Tariff Commission (1957) Synthetic Organic Chemicals, United States Production and Sales, 1956, Report No. 200, Second Series

US Tariff Commission (August 1971) Imports of Benzenoid Chemicals and Products, 1970, TC Publication 413

4,4'-METHYLENE BIS(2-METHYLANILINE)*

1. Chemical and Physical Data

1.1 Synonyms and trade names

Chem. Abstr. No.: None available

4,4'-Diamino-3,3'-dimethyldiphenylmethane; 3,3'-dimethyl-4,4'-di-
aminodiphenylmethane; 3,3'-dimethyldiphenylmethane-4,4'-diamine;
2,2'-dimethyl-4,4'-methylenedianiline; methylenebis(3-methylphenyl-
ene-4-amine); methylenebis(ortho-methylaniline); methylenebis
(ortho-toluidine); methylenebis(2-toluidine); 4,4'-methylenedi-
ortho-toluidine

1.2 Chemical formula and molecular weight

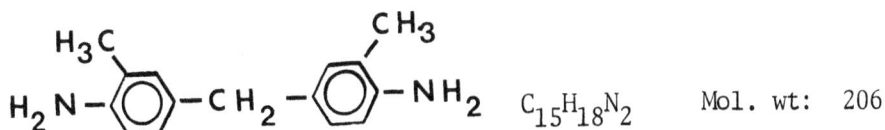

$H_2N-\bigcirc-CH_2-\bigcirc-NH_2$ with H_3C and CH_3 substituents $C_{15}H_{18}N_2$ Mol. wt: 206

1.3 Chemical and physical properties of the pure substance

(a) Description: White powder

(b) Melting-point: 149°C

(c) Solubility: Soluble in hot water, ethanol and probably in
other organic solvents

(d) Chemical reactivity: Weak base; has the general characteristics of primary aromatic amines

1.4 Technical products and impurities

No information is available to the Working Group on technical
products and impurities.

---

* Considered by the Working Group in Lyon, June 1973.

2. Production, Use, Occurrence and Analysis

(a) Production and use[1]

4,4'-Methylene bis(2-methylaniline) was produced commercially in the
United States in the past, probably by the reaction of formaldehyde with
ortho-toluidine. Apparently it was never produced for sale to others but
was used immediately as an unisolated intermediate in the manufacture of
the corresponding diisocyanate, 4,4'-methylene bis(ortho-tolylisocyanate).
Commercial production of this diisocyanate was first reported by one
company in 1955 (US Tariff Commission, 1956). A total of three companies
reported production of this diisocyanate at various times prior to 1964,
but no production has been reported since (US Tariff Commission, 1964).
All of these companies still manufacture diisocyanates for use in the manufacture of polyurethanes, but 4,4'-methylene bis(ortho-tolylisocyanate) is
no longer included among their products.

In 1972, one Japanese company was reported to be producing 4,4'-
methylene bis(2-methylaniline), but no information is available concerning
the quantity produced. No evidence was found that this diamine is
produced in other countries.

It is believed that 4,4'-methylene bis(2-methylaniline) has been
investigated experimentally for use as a curing agent for epoxy resins,
but no indication was found that it was ever used commercially for this
purpose.

4,4'-Methylene bis(2-methylaniline) can be used in the synthesis of
the dye C.I. Basic Violet 2 (The Society of Dyers and Colourists, 1971).
No information is available on the production of the dye in the US using
this substance, but it seems likely that the diamine has been used for the
manufacture of dyes in Japan, since the Japanese manufacturer of the
diamine also produces a line of dyes but no diisocyanates.

---

[1] Data from Chemical Information Services, Stanford Research
Institute, USA.

74

(b) Occurrence

4,4'-Methylene bis(2-methylaniline) has not been reported to occur as such in nature. It might be present in the waste streams from plants which produce it as an intermediate for further processing.

(c) Analysis

Guidelines for the analysis of aromatic amines are available (UICC, 1970). The reader is also referred to the literature cited in the Aniline Monograph, which gives methods for the separation and subsequent detection of a number of amines.

3. Biological Data Relevant to the Evaluation of Carcinogenic Risk to Man

3.1 Carcinogenicity and related studies in animals

(a) Oral administration

Rat: Munn (1967) administered 4,4'-methylene bis(2-methyl-aniline) to 24 male rats by gastric intubation 5 times per week as a solution in arachis oil for 10 months (total dose, 10.2 g/kg bw). Of 23 animals examined 18 had malignant liver tumours and 2 had benign liver tumours together with 12 subcutaneous tumours (mainly fibromas) within 487 days from the start of treatment. No controls were reported as having been used in the experiment.

Stula et al. (1971, 1973) reported that 50 male and 50 female rats administered 200 ppm 4,4'-dimethylene bis(2-methylaniline) in a standard diet (23% protein) for about one year developed tumours of the lung, liver and skin. The incidence of liver tumours, the frequency of malignant tumours and the number of metastases were higher in female as compared with male rats, but this might have been due to an age difference in survival. These tumours did not occur in the 50 male and 50 female control rats.

Another group of 25 male rats was administered 50 mg/kg bw of the compound in peanut oil by gavage for about 180 days; a group of 25 males served as controls. Survival was similar in both groups and reached a maximum of 16 months. Tumours of the liver, lung, mammary gland and skin appeared in treated rats. The liver tumour incidence was higher than that in the dietary feeding study, but the dose given was also 5 times as large (Stula et al., 1973).

3.2 Other relevant biological data

No data are available to the Working Group.

3.3 Observations in man

No data are available to the Working Group.

4. Comments on Data Reported and Evaluation[1]

4.1 Animal data

4,4'-Methylene bis(2-methylaniline) is carcinogenic in the rat after oral administration, the only species and route tested.

4.2 Human data

No epidemiological data are available to the Working Group.

----

[1] See also the section "Extrapolation from animals to man" in the introduction to this volume.

5. References

Munn, A. (1967) Occupational bladder tumors and carcinogens: recent developments in Britain. In: Deichmann, W. & Lampe, K.F., eds., Bladder Cancer. A Symposium, Birmingham, Alabama, Aesculapius, p. 187

The Society of Dyers and Colourists (1971) Colour Index, 3rd ed., 4, 4389

Stula, E.F., Sherman, H. & Zapp, J.A., Jr (1971) Experimental neoplasia in ChR-CD rats with the oral administration of 3,3'-dichlorobenzidine, 4,4'-methylene bis(2-chloroaniline), and 4,4'-methylene bis (2-methyl-aniline). Toxicol. appl. Pharmacol., 19, 380

Stula, E.F., Sherman, H., Zapp, J.A., Jr & Clayton, J.W., Jr (1973) Experimental neoplasia in rats from oral administration of 3,3'-di-chlorobenzidine, 4,4'-methylene-bis(2-chloroaniline) and 4,4'-methyl-ene-bis(2-methylaniline). Toxicol. appl. Pharmacol. (in press)

US Tariff Commission (1956) Synthetic Organic Chemicals, United States Production and Sales, 1955, Report No. 198, Second Series

US Tariff Commission (1964) Synthetic Organic Chemicals, United States Production and Sales, 1963, TC Publication 143

UICC (1970) The quantification of environmental carcinogens (UICC Technical Report Series, Vol. 4)

4,4'-METHYLENEDIANILINE*

1. Chemical and Physical Data

1.1 Synonyms and trade names

Chem. Abstr. No.: 101-77-9

Bis(p-aminophenyl)methane; bis(4-aminophenyl)methane; 4,4'-diamino-diphenylmethane; p,p'-diaminodiphenylmethane; dianilinomethane; 4,4'-methylenebisaniline; methylenedianiline; 4,4'-methylenedi-aniline; p,p'-methylenedianiline; DADPM; DAPM; MDA

1.2 Chemical formula and molecular weight

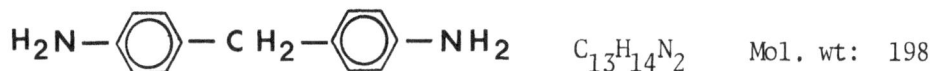

$$H_2N - \langle\bigcirc\rangle - CH_2 - \langle\bigcirc\rangle - NH_2 \qquad C_{13}H_{14}N_2 \qquad \text{Mol. wt: } 198$$

1.3 Chemical and physical properties of the pure substance

(a) Description: Pale yellow crystals which darken on standing in air

(b) Melting-point: $93^{\circ}C$

(c) Solubility: Slightly soluble in water; very soluble in alcohol, benzene and ether; soluble in many other organic solvents

(d) Stability: Oxidizes in air

(e) Chemical reactivity: Weak base; has the general characteristics of primary aromatic amines

1.4 Technical products and impurities

4,4'-Methylenedianiline (DAPM) is available as a commercial product in the form of flakes containing more than 99% active ingredient.

---

* Considered by the Working Group in Lyon, June 1973.

2. Production, Use, Occurrence and Analysis

(a) Production and use[1]

DAPM has been produced commercially for at least 50 years (US Tariff
Commission, 1922). The preferred commercial production method is the reaction of formaldehyde with aniline, which produces a mixture of di-, tri-
and polyamines from which the diamine can be isolated. However, only a
small amount of DAPM is isolated from this mixture, which is usually converted directly to polyisocyanates by reaction with phosgene. DAPM can
also be made by the reaction of p,p'-diaminobenzophenone with lithium
aluminium hydride, but this is a laboratory method.

Production of DAPM was reported by three United States companies in
1965 and 1966. Their combined production was approximately 500 thousand
kg in 1965 (US Tariff Commission, 1966) and about 700 thousand kg in 1966
(US Tariff Commission, 1967). In both years the total production was
understated, since one known producer was not included, and the quantities
reported did not include all the unisolated DAPM present in the mixture of
di-, tri- and polyamines used for polyisocyanate production. Data on
production and sales have not been published by the US Tariff Commission
since 1966, but it is estimated that annual US production of isolated DAPM
has been about one million kg in recent years. This quantity is much less
than that present in the intermediate amine mixture which is phosgenated
to yield the mixture of di-, tri- and polyisocyanates sold commercially
as polymethylene polyphenylisocyanate. In 1970, US production of
polymethylene polyphenylisocyanate is estimated to have been 52 million kg.
Although this mixed isocyanate is used primarily in the production of
polyurethane foams, some of the diisocyanate (4,4'-methylenediphenyl isocyanate) is separated from the mixed isocyanates and refined. US production of this refined diisocyanate is estimated to have been 5.5 million kg
in 1970 (primarily for use in the production of non-foamed polyurethanes).

---

[1] Data from Chemical Information Services, Stanford Research
Institute, USA.

US imports through the principal customs districts in 1970 were 2.7 thousand kg (US Tariff Commission, August 1971).

No data are available on the quantity of DAPM produced in countries other than the US. In 1967, the Federal Republic of Germany was reported to have one producer; the United Kingdom was reported to have three producers in 1970; and in 1972 Japan was reported to have two producers.

The mixture of di-, tri- and polyisocyanates and the diisocyanate from DAPM is made in Japan, the United Kingdom and the Federal Republic of Germany, and additional plants are expected to be in operation in the following countries by the end of 1973: Belgium, the German Democratic Republic, France, The Netherlands and Spain. If all of these plants start production, large quantities of DAPM will be required as an intermediate.

In the US the largest usage for unisolated DAPM is in the manufacture of polymethylene polyphenylisocyanate and 4,4'-methylenediphenyl isocyanate. The polymethylene polyphenylisocyanate is used in rigid polyurethane foam, and this usage is continuing to grow at a significant rate. The refined 4,4'-methylenediphenyl isocyanate is mostly used in the production of Spandex fibres, where the growth rate is also high.

The most important usage for isolated DAPM is believed to be in the production of the corresponding completely hydrogenated diamine, 4,4'-methylenebis(cyclohexylamine). In the past this hydrogenated diamine has been used almost exclusively to make 4,4'-methylene bis(cyclohexyl isocyanate). This saturated diisocyanate is used in light-stable, high-performance polyurethane coatings. In recent years, the hydrogenated diamine has also been used as an intermediate in the production of a new nylon yarn which is finding increasing use in clothing manufacture, and US production is estimated to have exceeded 3 million kg in 1972.

Isolated DAPM is also used: (i) as an intermediate in the production of poly(amide-imide) resins and fibres; (ii) as an intermediate in the synthesis of pararosaniline dye; and (iii) as a curing agent for liquid-castable polyurethane elastomers and epoxy resins. No information is available on the quantities consumed in these applications.

DAPM has been reported to be useful as a corrosion inhibitor and as a laboratory analytical reagent in the determination of tungsten and sulphates. It has a demonstrated activity as an anti-thyroid drug; however, it is not as potent as are derivatives of thiourea and does not seem to be of current interest for the treatment of hyperthyroidism.

(b) Occurrence

DAPM is not known to occur as such in nature. It could be present in the waste streams from plants where it is produced or used.

(c) Analysis

Guidelines for the analysis of aromatic amines are available (UICC, 1970). A description of the gas chromatographic identification of DAPM is also available (Li Gotti et al., 1970).

3. Biological Data Relevant to the Evaluation of Carcinogenic Risk to Man

3.1 Carcinogenicity and related studies in animals

(a) Oral administration

Rat: Munn (1967) administered DAPM at near the maximum tolerated dose by gastric intubation in arachis oil 5 days per week for 121 days (total dose, 3.3 g/kg bw). Of 24 male rats originally treated, 19 were alive at 12 months, 17 at 18 months and 12 at 2 years. All animals had cirrhosis of the liver, but no tumours were found before 2 years. Two benign hepatomas occurred at 792 and 947 days, as well as a variety of miscellaneous tumours.

In a second experiment, intragastric dosing was continued for about 18 months (total dose, 6 g/kg bw). Two liver tumours, 1 intestinal tumour, 1 pituitary tumour and 2 subcutaneous fibromas were found. The same strain was used for testing methylene bis-(2-methylaniline), which produced a high incidence of malignant liver tumours, showing that these rats were susceptible to liver carcinogens (Munn, 1967).

(b) Subcutaneous and/or intramuscular injection

Rat: Steinhoff & Grundmann (1970) injected groups of 25 male and 25 female Wistar rats with s.c. doses of 30-50 mg DAPM/kg bw in physiologic saline at 1 to 3 week intervals over a period of 700 days (average total dose, 1.4 g/kg bw). The mean survival time was 970 days in the males and 1060 days in the treated female animals, compared with 1007 days in the controls. A total of 29 benign and 33 malignant tumours was found in treated rats, compared with 15 benign and 16 malignant tumours in an equal number of controls. Four hepatomas were reported in the treated rats. The exact incidence of the tumour types observed was not recorded. In a later study using a similar number of rats, repeated s.c. injections over 410 days of 200-400 mg/kg bw (total dose, 7.3 g/kg bw) did not increase the incidence of benign and malignant tumours in treated rats compared with the incidence in controls. The mean survival time was reduced to 865 days in the treated group, compared with 1007 days in the controls.

3.2 Other relevant biological data

(a) Animals

When given as an i.p. injection in propylene glycol at a dose of 5 mg DAPM/kg bw, about 25% appears in the bile within 24 hours in the form of 3 or 4 unidentified metabolites. Kidney and liver damage is reflected in disturbed function tests (Smith & Millburn, 1966)[1].

(b) Man

The severe hepatotoxic effects of exposure to DAPM are well documented by Kopelman et al. (1966a) in their description of 84 persons affected in the "Epping Jaundice" outbreak, which followed the accidental contamination of flour used for baking bread. Hepatocellular

---

[1] Personal communication (cited by Kopelman et al., 1966b).

damage was mirrored by raised SGOT and SGPT levels but normal serum alkaline phosphatase (Kopelman et al., 1966b).

3.3 Observations in man

No epidemiological studies are available to the Working Group.

4. Comments on Data Reported and Evaluation[1]

4.1 Animal data

The available experimental evidence in the rat, the only species tested, does not permit a definite conclusion regarding the carcinogenicity of 4,4'-methylenedianiline (DAPM) in this species.

4.2 Human data

No epidemiological data are available to the Working Group.

---

[1] See also the section "Extrapolation from animals to man" in the introduction to this volume.

5. References

Kopelman, H., Robertson, M.H., Sanders, P.G. & Ash, I. (1966a) The Epping
    jaundice. Brit. med. J., i, 514

Kopelman, H., Scheuer, P.J. & Williams, R. (1966b) The liver lesion of
    the Epping jaundice. Quart. J. Med., 35, 553

Li Gotti, I., Piacentini, R. & Bonomi, G. (1970) Metodo rapido per la
    identificazione dei costituenti dei poliuretani da poliestere e delle
    resine poliestere non modificate. Mater. Plast. Elastomeri, 36, 213

Munn, A. (1967) Occupational bladder tumors and carcinogens: recent
    developments in Britain. In: Deichmann, W. & Lampe, K.F., eds.,
    Bladder Cancer. A Symposium, Birmingham, Alabama, Aesculapius, p. 187

Steinhoff, D. & Grundmann, E. (1970) Zur cancerogenen Wirkung von 4,4'-Di-
    aminodiphenylmethan und 2,4'-Diaminodiphenylmethan.
    Naturwissenschaften, 57, 247

UICC (1970) The quantification of environmental carcinogens (UICC
    Technical Report Series, Vol. 4)

US Tariff Commission (1922) Census of Dyes and other Synthetic Organic
    Chemicals 1921, Tariff Information Series No. 26

US Tariff Commission (1966) Synthetic Organic Chemicals, United States
    Production and Sales, 1965, TC Publication 206

US Tariff Commission (1967) Synthetic Organic Chemicals, United States
    Production and Sales, 1966, TC Publication 248

US Tariff Commission (August 1971) Imports of Benzenoid Chemicals and
    Products, 1970, TC Publication 413

1-NAPHTHYLAMINE[*]

1. Chemical and Physical Data

1.1 Synonyms and trade names

Chem. Abstr. No.: 134-32-7

Alpha-naphthylamine; 1-aminonaphthalene
C.I. Azoic Diazo Component 114 (Colour Index); Fast Garnet B Base;
Fast Garnet Base B; Naphthalidam; Naphthalidine

1.2 Chemical formula and molecular weight

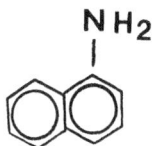

$C_{10}H_9N$ Mol. wt: 143.2

1.3 Chemical and physical properties of the pure substance

(a) Description: Colourless crystals which darken in air to purple
red colour; unpleasant odour

(b) Boiling-point: $301^\circ C$ ($104^\circ C$ at 10 mm)

(c) Melting-point: $50^\circ C$

(d) Solubility: At $25^\circ C$, 0.16% dissolves in water; freely soluble
in alcohol and ether and many other organic solvents

(e) Stability: Oxidizes in air

(f) Volatility: Sublimes, and is volatile in steam

(g) Chemical reactivity: A weak base; has the general characteristics of primary aromatic amines

---

[*] Considered by the Working Group in Lyon, June 1973.

1.4 Technical products and impurities

Most commercial 1-naphthylamine, which is prepared by nitration of
naphthalene and reduction of the products, contains 4-10% of 2-naphthyl-
amine if synthesized by methods used in previous decades. Modern production methods give rise to a maximum of 0.5%.

2. Production, Use, Occurrence and Analysis

(a) Production and use[1]

1-Naphthylamine has been produced commercially for at least 50 years
(US Tariff Commission, 1922). Although it was produced from 1-nitronaph-
thalene by reduction with iron and hydrochloric acid in the past, in
recent years it is believed to have been made exclusively by catalytic
hydrogenation of 1-nitronaphthalene using a nickel catalyst.

Total sales in the United States were last reported in 1949 when they
amounted to 255 thousand kg (US Tariff Commission, 1950). Total US sales
in 1948 were reported to have been 567 thousand kg, and total production
was reported to have been 2.6 million kg (US Tariff Commission, 1949).
In 1963, one source estimated US production of 1-naphthylamine at more
than 500 thousand kg (Shreve, 1963). By 1972, only one US company was
still producing 1-naphthylamine.

US imports through the principal customs districts have grown
steadily from the 2 thousand kg reported for 1967 (US Tariff Commission,
September 1968) and reached 27 thousand kg in 1971 (US Tariff Commission,
July 1972).

No data are available on the quantity of 1-naphthylamine produced in
countries other than the US. In 1967 the Federal Republic of Germany was
reported to have two producers; Italy was reported to have three producers of naphthylamine and derivatives (not further specified) in 1969;
in 1970 the United Kingdom was reported to have two producers; and Japan

---

[1] Data from Chemical Information Services, Stanford Research
Institute, USA.

was reported to have three producers in 1972.

1-Naphthylamine is used as a chemical intermediate in the preparation of a large number of compounds. The major use appears to be in the manufacture of dyes, but the manufacture of herbicides and antioxidants are also believed to be significant outlets.

1-Naphthylamine can be used as an intermediate in the production of 150 dyes (The Society of Dyers and Colourists, 1971). Another source has estimated that it is actually used as an intermediate in the preparation of more than 50 dyes (Anon., 1973). It is known that 1-naphthylamine is used as an intermediate in the manufacture of 1-naphthylamine-4-sulphonic acid (naphthionic acid); this acid is used to make 1-naphthol-4-sulphonic acid (Neville-Winther's acid - a coupling agent for azo dyes) and its sodium salt (which serves as a dye intermediate). Total US production of naphthionic acid was last reported for the year 1969 when it amounted to 81 thousand kg (US Tariff Commission, 1971).

1-Naphthylamine is used to manufacture the herbicide N-1-naphthyl-phthalamic acid. No production figures are available for this compound, but the results of one survey indicated that US farm usage of 1-naphthyl-phthalamic acid on all crops amounted to an estimated 450 thousand kg in 1966 (US Department of Agriculture, April 1970). As of August 31, 1968 the acid or its sodium salt were approved in the US for use on five fruits, eight vegetables and on cotton and peanuts (US Department of Agriculture, 1968).

Rubber antioxidants produced from 1-naphthylamine include N-phenyl-1-naphthylamine, and the aldol-1-naphthylamine condensate. Data on production volumes are not available, but it is believed that these antioxidants are less significant commercially than are N-phenyl-2-naphthylamine and derivatives; in any case, all of the N-phenyl-1-naphthylamine may not be made from 1-naphthylamine, since it can also be made by the condensation of aniline with 1-naphthol.

It has been reported that 1-naphthylamine has also been used for the manufacture of 1-naphthol, but this is not believed to be commercially significant. 1-Naphthylamine is believed to have been used in the production of a rodenticide, 1-naphthylthiourea (antu), but this chemical is no longer produced in the US.

(b) Occurrence

1-Naphthylamine has not been reported to occur as such in nature, but it has been found to be present in coal-tar (Treibl, 1967). It may also be present in the waste streams from plants where it is produced or used. Cigarette smoke contains 0.03 µg of 1-naphthylamine per cigarette (Masuda & Hoffmann, 1969).

1-Naphthylamine (containing less than 1% of 2-naphthylamine as a by-product of a chemical reaction) is a controlled substance under the UK Carcinogenic Substances Regulations 1967 - Statutory Instrument (1967) No. 879. It is also included with 13 other compounds in the US Federal Register (US Government, 1973) as being subject to an Emergency Temporary Standard on certain carcinogens under an order made by the Occupational Safety and Health Administration, Department of Labor, on 26 April, 1973.

(c) Analysis

Guidelines for the analysis of aromatic amines are available (UICC, 1970). Specific information on the detection of 1-naphthylamine is given in a paper by Shimomura & Walton (1968) in which separation from other amines is carried out using thin-layer chromatography. A gas chromatographic method, in which 1- and 2-naphthylamines are separated after reaction with pentafluoropripionic anhydride, will detect as little as 1 mg (Masuda & Hoffmann, 1969).

3. Biological Data Relevant to the Evaluation of Carcinogenic Risk to Man

3.1 Carcinogenicity and related studies in animals

(a) Oral administration

Mouse: Clayson & Ashton (1963) administered solutions of 0.01% 1-naphthylamine hydrochloride (freed from 2-naphthylamine by multiple fractional recrystallizations) in the drinking-water of 61 male and female stock mice over a period of 84 weeks. In the males the incidence of hepatomas was 4/18 (22%) in treated animals and 4/24 (17%) in the controls. In females the corresponding incidences were 5/43 (11%) and 0/36 (0%).

Hamster: 1-Naphthylamine administered in the diet at concentrations of 0.1% for life and 1% for 70 weeks to 2 groups of 30 male and 30 female Syrian golden hamsters failed to show a carcinogenic effect (Saffiotti et al., 1967; Sellakumar et al., 1969).

Dog: Doses of 0.5 g 1-naphthylamine were given to 2 mongrel dogs 3 times per week for life. After 9 years 1 dog had a bladder papilloma (Bonser et al., 1958). 1-Naphthylamine of two levels of purity, neither freed from the 2-isomer, was given to 5 dogs in gelatine capsules in 300 or 330 mg doses 5 times weekly for $4\frac{1}{2}$ years without inducing bladder tumours (Gehrmann et al., 1949).

(b) Subcutaneous and/or intramuscular administration

Newborn mouse: In male and female Swiss mice given 100 µg 1-naphthylamine/dose s.c. on the 1st, 3rd and 5th days of life, 5/35 male mice had tumours after 12 months when the animals were sacrificed. Findings included 3 pulmonary tumours, 1 hepatoma and 1 lymphosarcoma. In control mice only 1 lymphosarcoma was observed. One lung adenoma was found in the treated female mice, with no tumours in the female controls. A single injection of 30 µg on the first day of life resulted in the occurrence of 4 tumours in 65 treated mice (3 lung tumours and 1 hepatoma) at 10 months. No tumours were observed in the controls (Radomski et al., 1971).

3.2 Other relevant biological data

(a) Animals

The metabolis of 1-naphthylamine has not been so fully studied as that of 2-naphthylamine. Clayson & Ashton (1963) demonstrated by paper chromatography that 1-amino-2-naphthol and 1-amino-4-naphthol conjugates are formed in several species. Brill & Radomski (1967) and Deichmann & Radomski (1969) found that N-(1-naphthyl)-hydroxylamine was present in smaller concentrations than was N-(2-naphthyl)-

hydroxylamine after administration of the appropriate naphthylamine
to the dog, and that no 1-nitrosonaphthalene was formed. Radomski
et al.(1971) found that significant amounts of N-(1-naphthyl)-hydroxy-
lamine and 1-nitrosonaphthalene were excreted only when a high dose
of 1-naphthylamine was administered to dogs. This contrasts with the
behaviour of 2-naphthylamine. The diphosphate ester of 1-amino-2-
naphthol has not been identified in the urine of dogs (Deichmann &
Radomski, 1969).

(b) Man

Doses of 250 mg 1-naphthylamine given to 4 patients resulted in
the excretion of the N-hydroxy-1-naphthylacetamide either in the free
state or as a glucuronic acid conjugate (Belman et al., 1968).

(c) Carcinogenicity of metabolites

The carcinogenicity of 1-amino-2-naphthol hydrochloride following
bladder implantation in paraffin pellets in mice was reported by
Bonser et al.(1963).

Boyland et al.(1964), using bladder implantation, obtained
bladder tumours in 8/26 mice when N-(1-naphthyl)-hydroxylamine was
implanted in a 20% suspension in stearic acid pellets. Stearic acid
pellets alone produced bladder tumours in 8/62 controls surviving up
to 40 weeks.

Radomski et al.(1971) tested N-(1-naphthyl)-hydroxylamine in
Osborne-Mendel rats by i.p. injection at 100 mg/kg bw/week in corn oil
for 12 weeks and obtained tumours at various sites in 12/27 animals
with a median survival time of 26 weeks, as compared to a single
tumour in 1/27 untreated animals surviving up to 40 weeks. Similar
i.p. injections of 1-nitrosonaphthalene (but in a lower amount due to
the hepatotoxicity of the compound) produced tumours in 4/27 treated
animals surviving for 28 weeks. One tumour occurred in 1/27 control
animals surviving for 40 weeks.

3.3 Observations in man

(a) Case reports

Cases of bladder tumours attributed to 1-naphthylamine have been reported by a number of workers (Wignall, 1929; Evans, 1937; Gehrmann et al., 1949; Di Maio, 1949; Barsotti, 1949; Billiard-Duchesne, 1949; Goldblatt, 1949; Scott, 1952).

(b) Epidemiology

Case et al. (1954) showed that 5 years' exposure to commercial 1-naphthylamine was necessary for the induction of bladder tumours. In their survey, they found 6 deaths certificated as due to bladder cancer among 1-naphthylamine workers who had not also been engaged in the production of 2-naphthylamine or benzidine, whereas only 0.70 would have been expected ($P<0.005$). The morbidity was 19 cases.

4. Comments on Data Reported and Evaluation

4.1 Animal data

No carcinogenic effect of 1-naphthylamine was observed in the hamster following oral administration. The results obtained after oral and subcutaneous administration to mice are inconclusive. The experiments in dogs demonstrate that 1-naphthylamine, if carcinogenic at all, is less so to the bladder in this species than is the 2-isomer.

4.2 Human data

Occupational exposure to commercial 1-naphthylamine containing 4-10% 2-naphthylamine is strongly associated with bladder cancer. A number of case reports from several countries support this association. It is not possible on present evidence to decide whether 1-naphthylamine free from the 2-isomer is carcinogenic to man.

5. References

Anon. (1973) Market Newsletters. Chemical Week, May 23, 31

Barsotti, M. & Vigliani, E.C. (1949) Lesioni vescicali da amine aromatiche. Med.d.Lavoro, 40, 129

Belman, S., Troll, W., Teebor, G. & Mukai, F. (1968) The carcinogenic and mutagenic properties of N-hydroxy-aminonaphthalenes. Cancer Res., 28, 535

Billiard-Duchesne, J.F. (1949) Les amino-tumeurs de la vessie en France. In: Proceedings of the Ninth International Congress on Industrial Medicine, London, 1948, Bristol, Wright. p. 507

Bonser, G.M., Boyland, E., Busby, E.R., Clayson, D.B., Grover, P.L. & Jull, J.W. (1963) A further study of bladder implantation in the mouse as a means of detecting carcinogenic activity - use of crushed paraffin wax or stearic acid as the vehicle. Brit. J. Cancer, 17, 127

Bonser, G.M., Clayson, D.B. & Jull, J.W. (1958) Some'aspects of the experimental induction of tumours of the bladder. Brit. med. Bull., 14, 146

Boyland, E., Busby, E.R., Dukes, C.E., Grover, P.L. & Manson, D. (1964) Further experiments on implantation of materials into the urinary bladder of mice. Brit. J. Cancer, 18, 575

Brill, E. & Radomski, J. (1967) N-hydroxylation of 1-naphthylamine in the dog. Life Sci., 6, 2293

Case, R.A.M., Hosker, M.E., McDonald, D.B. & Pearson, J.T. (1954) Tumours of the urinary bladder in workmen engaged in the manufacture and use of certain dyestuff intermediates in the British chemical industry. I. The role of aniline, benzidine, alpha-naphthylamine and beta-naphthylamine. Brit. J. industr. Med., 11, 75

Clayson, D.B. & Ashton, M.J. (1963) The metabolism of 1-naphthylamine and its bearing on the mode of carcinogenesis of the aromatic amines. Acta Un. int. Cancr., 19, 539

Deichmann, W.B. & Radomski, J. (1969) Carcinogenicity and metabolism of aromatic amines in the dog. J. nat. Cancer Inst., 43, 263

Di Maio, G. (1949) Affections of the bladder due to aromatic amines. In: Proceedings of the Ninth International Congress on Industrial Medicine, London, 1948, Bristol, Wright. p. 476

Evans, E.E. (1937) Causative agents and protective measures in aniline tumour of the bladder. J. Urol., 38, 212

Gehrmann, G.H., Foulger, J.H. & Fleming, A.J. (1949) Occupational carcinoma of the bladder. In: Proceedings of the Ninth International Congress on Industrial Medicine, London, 1948, Bristol, Wright. p. 472

Goldblatt, M.W. (1949) Acute haemorrhagic cystitis and vesical tumours induced by chemical compounds in industry. In: Proceedings of the Ninth International Congress on Industrial Medicine, London, 1948, Bristol, Wright. p. 497

94

Masuda, Y. & Hoffmann, D. (1969) Quantitative determination of 1-naphthyl-amine and 2-naphthylamine in cigarette smoke. Analyt. Chem., 41, 650

Radomski, J.L., Brill, E., Deichmann, W.B. & Glass, E.M. (1971) Carcino-genicity testing of N-hydroxy and other oxidation and decomposition products of 1- and 2-naphthylamine. Cancer Res., 31, 1461

Saffiotti, U., Cefis, F., Montesano, R. & Sellakumar, A.R. (1967) Induc-tion of bladder cancer in hamsters fed aromatic amines. In: Deichmann, W. & Lampe, K.F., eds., Bladder Cancer. A Symposium, Birmingham, Alabama, Aesculapius, p. 129

Scott, T.S. (1952) The incidence of bladder tumours in a dyestuffs factory. Brit. J. industr. Med., 9, 127

Sellakumar, A.R., Montesano, R. & Saffiotti, U. (1969) Aromatic amines carcinogenicity in hamsters. Proc. amer. Assoc. Cancer Res., 10, 78

Shreve, R.N. (1963) Amination by reduction. In: Kirk, R.E. & Othmer, D.F., eds., Encyclopedia of Chemical Technology, 2nd ed., New York. John Wiley & Sons, Vol. 2, p. 82

Shimomura, K. & Walton, H.F. (1968) Thin-layer chromatography of amines by ligant exchange. Separation Sci., 3, 493

The Society of Dyers and Colourists (1971) Colour Index, 3rd ed., 4, 4806

Treibl, H.G. (1967) Naphthalene derivatives. In: Kirk, R.E. & Othmer, D.F., eds., Encyclopedia of Chemical Technology, 2nd ed., New York. John Wiley & Sons, Vol. 13, p. 707

UICC (1970) The quantification of environmental carcinogens (UICC Technical Report Series, Vol. 4)

US Department of Agriculture (1968) Summary of Registered Agricultural Pesticide Chemical Uses, I, I-N-4.1-I-N-5.2

US Department of Agriculture (April 1970) Economic Research Service, Quantities of Pesticides Used by Farmers in 1966, Agricultural Economic Report No. 179

US Government (1973) Occupational safety and health standards. US Federal Register, 38, No. 85, 10929

US Tariff Commission (1922) Census of Dyes and other Synthetic Organic Chemicals, 1921, Tariff Information Series No. 26

US Tariff Commission (1949) Synthetic Organic Chemicals, United States Production and Sales, 1948, Report No. 164, Second Series

US Tariff Commission (1950) Synthetic Organic Chemicals, United States Production and Sales, 1949, Report No.169, Second Series

US Tariff Commission (September 1968) Imports of Benzenoid Chemicals and Products, 1967, TC Publication 264

US Tariff Commission (1971) Synthetic Organic Chemicals, United States Production and Sales, 1969, TC Publication 412

US Tariff Commission (July 1972) Imports of Benzenoid Chemicals and Products 1971, TC Publication 466

Wignall, T.H. (1929) Incidence of disease of the bladder in workers in certain chemicals. Brit. med. J., ii, 258

2-NAPHTHYLAMINE[*]

1. Chemical and Physical Data

1.1 Synonyms and trade names

Chem. Abstr. No.: 91-59-8

2-Aminonaphthalene; beta-naphthylamine; BNA; Fast Scarlet Base B

1.2 Chemical formula and molecular weight

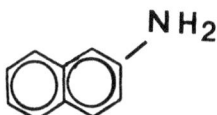

$C_{10}H_9N$ Mol. wt: 143.2

1.3 Chemical and physical properties of the pure substance

(a) Description: Colourless crystals which darken in air to a reddish-purple colour

(b) Boiling-point: $306^{\circ}C$ ($158^{\circ}C$ at 10 mm)

(c) Melting-point: $111-113^{\circ}C$

(d) Solubility: Soluble in hot water, alcohol, ether and many other organic solvents

(e) Stability: Stable in cold in the absence of air; oxidizes in the presence of air

(f) Volatility: Volatile in steam

(g) Chemical reactivity: A weak base; has the general characteristics of primary aromatic amines

---

[*] Considered by the Working Group in Lyon, June 1973.

1.4 Technical products and impurities

The commercial product contains a number of polyaromatic heterocyclic compounds including 3,4,5,6-dibenzophenazine, which is formed from 2-naphthylamine in the presence of air. 2-Amino-1,4-naphthoquinone-N^4,2-naphthylimine has also been reported as a contaminant of 2-naphthylamine (Radomski et al., 1971).

2. Production, Use, Occurrence and Analysis

(a) Production and use[1]

2-Naphthylamine has been produced commercially for at least 50 years (US Tariff Commission, 1922). The usual method of commercial manufacture is believed to have been the reaction of 2-naphthol with ammonia and ammonium sulphite.

Data for United States production in 1955 (the latest available) indicate that a total of 581 thousand kg were produced by four manufacturers (US Tariff Commission, 1956). Commercial production of 2-naphthylamine reportedly continued in the US until early 1972 (Hyatt, 1973), but production in recent years is believed to have been quite small.

US imports through the principal customs districts were last of significance in 1967 when they amounted to 17.4 thousand kg (US Tariff Commission, September 1968).

Data on production of 2-naphthylamine in countries other than the US are not available. The United Kingdom and several other countries reportedly discontinued production prior to 1960 because of evidence of bladder cancer resulting from exposure to the chemical (Kilner & Samuel, 1960). In 1969, the Federal Republic of Germany was reported to have had one producer of "naphthylamine and derivatives" (not further specified) in 1967; and Japan was reported in 1972 to have three producers.

---

[1]Data from Chemical Information Services, Stanford Research Institute, USA.

98

The major use for 2-naphthylamine appears to have been as an intermediate in the manufacture of dyes and antioxidants.

2-Naphthylamine can be used as an intermediate in the production of twenty-one dyes. However, it is not now usually used as such, and frequently 2-amino-1-naphthalene-sulphonic acid is substituted for it (The Society of Dyers and Colourists, 1971).

At one time 2-naphthylamine was the preferred intermediate for the manufacture of J-acid (2-amino-5-naphthol-7-sulphonic acid) and gamma-acid (7-amino-1-naphthol-3-sulphonic acid), but these two dye intermediates are now made by alternative methods (Kilner & Samuel, 1960). It is believed that 2-naphthylamine was used until recent years in the synthesis of N-alkyl-2-naphthylamines (e.g., N-ethyl-2-naphthylamine), which found use as dye intermediates.

Although the antioxidant N-phenyl-2-naphthylamine and another antioxidant derived from it, N-phenyl-2-naphthylamine-acetone condensate, may have been made from 2-naphthylamine in the US in the past, it is probable that these products have been manufactured by the condensation of aniline with 2-naphthol in recent years. N-Phenyl-2-naphthylamine is known to be produced in countries other than the US, but the method of synthesis is not known.

(b) Occurrence

2-Naphthylamine has not been reported to occur as such in nature. However, it is formed in the pyrolysis of nitrogen-containing organic matter. For example, the pyrolysis of L-glutamic acid and L-leucine at 700°C (but not at 500°C) may produce 1- and 2-naphthylamine (Masuda et al., 1967). 2-Naphthylamine reportedly has been found in coal-tar (Treibl, 1967); gas retort houses have been shown to have atmospheric levels of 2-naphthylamine sufficient to give a human exposure of 0.2 μg per day (Battye, 1966). Cigarette smoke contains 0.02 μg of 2-naphthylamine per cigarette, which is equivalent to 1 μg per 50 cigarettes, the daily exposure of a heavy cigarette smoker (Hoffmann et al., 1969).

2-Naphthylamine is present as an impurity in commercial 1-naphthylamine. The specifications for US-produced 1-naphthylamine indicate that the content is at present 0.5% or less. Other commercial 1-naphthylamine may contain higher percentages. 2-Naphthylamine may occur in the waste streams from plants where it is produced and used; it has been reported to be present in the effluent from certain dyestuffs factories in Japan (Takemura et al., 1965).

The presence, use and manufacture of β-naphthylamine in factories is prohibited in the UK under the Carcinogenic Substances Regulations 1967 - Statutory Instrument (1967) No. 487. It is also listed with 13 other compounds in the US Federal Register (US Government, 1973) as subject to an emergency temporary standard on certain carcinogens under an order made by the Occupational Safety and Health Administration, Department of Labor, on 26 April, 1973.

(c) Analysis

Guidelines for the analysis of aromatic amines are available (UICC, 1970).

Methods to determine relatively gross amounts of 2-naphthylamine in the presence of other industrial aromatic amines have been described (Butt & Strafford, 1956). A colorimetric determination of the oxidation products obtained from 2-naphthylamine and a large number of other aromatic amines has been published (Gupta & Srivastava, 1971).

2-Naphthylamine can be separated from other non-volatile aromatic amines by thin-layer chromatography (Sawicki et al., 1966). Separation from 1-naphthylamine can also be achieved by thin-layer chromatography (Shimomura & Walton, 1968; Ghetti et al., 1968). A gas chromatographic method in which 1- and 2-naphthylamine are separated after reaction with pentafluoropropionic anhydride will detect levels down to 1 ng (Masuda & Hoffmann, 1969).

3. Biological Data Relevant to the Evaluation
of Carcinogenic Risk to Man

3.1 Carcinogenicity and related studies in animals

(a) Oral administration

Mouse: In groups of 23 DBA and 25 IF mice given 2-naphthylamine by stomach tube at dosages of 240 and 400 mg/kg bw/week in arachis oil, respectively, for 90 weeks, liver tumours developed in 50% of the animals. In 2 similar groups of control mice given arachis oil alone no hepatomas were found, although an incidence of 8% occurred in control breeding stock. Dietary administration of 160 mg/kg bw/week to 4 groups of DBA mice fed 4 different synthetic diets resulted in the production of malignant hepatomas in 16/111 mice (Bonser et al., 1952).

Rat: Feeding of 0.01% 2-naphthylamine in the diet to albino Glaxo rats for life led to 4 possibly pre-cancerous changes of the bladder in 31 rats surviving for up to 102 weeks, but not to carcinomas (Bonser et al., 1952). Oral administration by gavage of 0.5-150 mg per week to rats for 52 weeks induced no malignant changes in a total of 59 rats which were observed only up to 80 weeks (Hadidian et al., 1968).

Hamster: Lifetime administration of 1% 2-naphthylamine in the diet of 30 female and 30 male Syrian golden hamsters produced bladder carcinomas in 50% of the animals, the first tumours appearing after 45 weeks. One male and 1 female animal **also** developed hepatomas. Dietary concentrations of 0.1% failed to evoke carcinomas in 60 animals after 97 weeks (Saffiotti et al., 1967; Sellakumar et al., 1969).

Rabbit: Doses of 400 mg per week for up to 5 years induced papillomas of the bladder in only 1 out of 6 rabbits (Bonser et al., 1952).

Dog: In a preliminary report, Hueper et al.(1938) reported that commercial 2-naphthylamine given mainly by mouth but also, in part, by s.c. injection for more than 90 weeks induced 2 papillomas and 6 carcinomas of the bladder in 16 dogs. This evidence was based partly on autopsy and partly on cystoscopy and biopsy - the final results were not published. Bonser (1943) reported the occurrence of multiple papillomas in 3 out of 4 dogs given partially purified 2-naphthylamine in doses of 100-700 mg/day for 3 to 5 years. One died after 1 year of treatment. Later it was shown by Bonser et al.(1956) that 2-naphthylamine rigorously purified by gradient sublimation and fed orally to 4 female dogs in doses of 200-600 mg/day on 6 days per week for life resulted in the occurrence of multiple transitional cell papillomas of the bladder in 2 dogs which survived for 2 years or more. In 1 of these dogs which was killed after 3 years an invasive carcinoma was present. Conzelman & Moulton (1972) fed rigorously purified 2-naphthylamine in doses of 37.5, 75, 150 and 300 mg/kg bw/week for 6 to 26 months to both male and female beagle dogs and obtained carcinomas of the bladder at each of the 4 levels used within 30 months. Carcinomas were present in 9/11 dogs which received a total of 100-200 g 2-naphthylamine, whereas only 6/22 occurred in dogs receiving total doses of less than 100 g.

Monkey: Conzelman et al.(1969) administered by stomach tube increasing amounts of 2-naphthylamine (from 37.5-2,400 mg/kg bw/week) to groups of rhesus monkeys for periods of up to 60 months. Of 24 animals, 3 developed papillary adenomas and 9 transitional cell carcinomas of the bladder between 143 and 256 weeks after the beginning of treatment. The majority of tumours occurred in animals given high doses of 2-naphthylamine. No other types of tumours were observed, and no tumours occurred in 3 control animals.

(b) Subcutaneous and/or intramuscular injection

Mouse: Bonser et al.(1956) gave repeated s.c. injections of 2-naphthylamine to stock and CBA mice. Hepatomas and a variable incidence of local sarcomas were obtained. The incidence of sarcomas was 10/16 in mice surviving 20 weeks or more; and 4/5 mice surviving for more than 80 weeks had hepatomas when commercial 2-naphthylamine was allowed to age in the arachis oil solution before injection. A lower incidence, 2/12 mice with sarcomas and 1/12 mice with hepatomas, was obtained when the rigorously purified solution was prepared freshly prior to injection. No sarcomas (0/10) were found in mice surviving 58 weeks or more when an aqueous solution of the rigorously purified hydrochloride was used, nor when the commercial naphthylamine was freshly prepared in oil (0/13); but hepatomas occurred in 2/8 mice surviving more than 70 weeks and in 2/4 mice surviving 77 weeks, respectively. The different effect obtained when the solution of 2-naphthylamine was allowed to age may possibly be due to the formation of carcinogenic products through oxidation.

Newborn mouse: S.c. injections of 100 μg 2-naphthylamine in 3% aqueous gelatine into BALB/c mice once daily for the first 5 days of life resulted in a 21.9% incidence of lung adenomas versus 8.3% in control animals, all survivors being killed after 50 weeks. The s.c. injection of 100 μg 2-naphthylamine in arachis oil either once on the first day of life or once daily for the first 5 days of life to BALB/c mice gave negative results. In this case the duration of the experiment was 40 weeks (Walters et al., 1967).

(c) Other experimental systems

Intravesicular instillation: Instillation of 10 mg 2-naphthylamine dissolved in dimethylsulphoxide once every 2 weeks for 30 months into the bladder lumina of 4 dogs did not induce tumours of the bladder or other tissues when the dogs were sacrificed at 45 months (Radomski et al., 1971).

3.2 Other relevant biological data

(a) Animals

Boyland (1958) and Boyland & Manson (1966) identified 24 metabolites in the urine of rats, rabbits, dogs or monkeys. The following mechanisms represent the metabolic pathways:-

(i) N-hydroxylation followed by conversion to 2-amino-1-naphthyl mercapturic acid, 2-nitrosonaphthalene and rearrangement to 2-amino-1-naphthol

(ii) Oxidation at C_5 and C_6 to an arene oxide which rearranges to 5-hydroxy-2-naphthylamine, reacts with water to form a 5,6-dihydroxy dihydro derivative and forms a 5-hydroxy-6-mercapturic acid

(iii) Conjugation of the amino group with acetic, sulphuric or glucosiduronic acid

103

(iv) Secondary conjugation of the hydroxyl group with phosphate
      sulphuric or glucosiduronic acid

The proportion of these metabolites excreted in the urine of
various experimental animals has been shown to differ (Bonser et al.,
1951; Boyland & Manson, 1966; Conzelman et al., 1969), although in
most species 2-amino-1-naphthyl sulphate is the predominant metabolite. In early studies it was thought that the ortho-hydroxylation
metabolite, 2-amino-1-naphthyl glucosiduronic acid, may be hydrolyzed
by β-glucuronidase present in urine yielding 2-amino-1-naphthol which
may act as the proximate carcinogen (Allen et al., 1957).

In dogs a small percentage of 2-naphthylamine was metabolized to
bis-(2-amino-1-naphthyl) hydrogen phosphate (Boyland et al., 1961;
Troll et al., 1963). N-Oxidation of 2-naphthylamine leads to more
reactive metabolites (Deichmann & Radomski, 1969). Radomski & Brill
(1971) showed that after a single oral dose of 70 mg/kg bw of both
the carcinogenic 2-isomer and the weakly active, or inactive,
1-isomer the proportion of the dose converted into the corresponding
N-hydroxylamine and nitroso compound and excreted in the urine was
about the same. When a dose of 5 mg/kg bw was given, the proportion
of the 2-isomer converted to these metabolites in the urine remained
about the same (0.2%), whereas the 1-isomer gave rise to barely
detectable traces of these compounds.

The unstable N-hydroxy metabolites are excreted as glucuronic
acid conjugates and are hydrolyzed by acid or β-glucuronidase.
Radomski et al.(1973) consider these conjugates to be the carcinogenic urinary metabolites of aromatic amines in the dog. Methaemoglobinaemia, which is a measure of N-hydroxy compounds in the blood,
was higher in the dog with the 2-isomer at the 70 mg/kg dose
(Radomski & Brill, 1971).

(b) Man

Because of the known carcinogenicity of 2-naphthylamine in man,
few metabolic studies have been carried out. However, N-(2-naphthyl)
-hydroxyl-amine and bis-(2-amino-1-naphthyl) phosphate have been

identified in the urine of some hospital patients given small doses of 2-naphthylamine (Troll & Nelson, 1961; Troll et al., 1963).

(c) Carcinogenicity of metabolites

N-(2-Naphthyl)-hydroxylamine was shown to be carcinogenic by bladder implantation in mice (Bonser et al., 1963) and by i.p. injection in rats (Boyland et al., 1963).

Radomski et al.(1971) compared the carcinogenicity of N-(1- and 2-naphthyl)-hydroxylamine and 1- and 2-nitrosonaphthalene administered to newborn mice and adult rats by s.c. and i.p. routes, respectively. In groups of male and female Swiss mice given three s.c. injections of 100 μg of the respective chemicals on the first, third and fifth days of life, a significant increase in the incidence of lung adenomas in females and of hepatomas in males was observed after 1 year in mice given N-(2-naphthyl)-hydroxylamine. A lower incidence of lung adenomas was observed after treatment with N-(1-naphthyl)-hydroxylamine and 1- and 2-nitrosonaphthalene than with N-(2-naphthyl)-hydroxylamine. The administration of 2-nitrosonaphthalene resulted, however, in a high incidence of hepatomas in male mice. No lung or liver tumours were seen in control animals.

The i.p. injections of N-(1-naphthyl)-hydroxylamine to male Osborne-Mendel rats resulted in a significant increase in the number of tumours occurring at various sites (12/27 in the treated animals versus 1/27 in the controls). The i.p. injection of 1-nitrosonaphthalene resulted in a barely significant increase in the number of tumours occurring at various sites (4/27 in treated versus 1/27 in controls). The administration of the 2-isomers gave completely negative results. It must be noted, however, that the life span of the rats was severely reduced due to the hepatotoxicity of these compounds (Radomski et al., 1971).

An s.c. injection into groups of 50 to 60 BALB/c mice of 100 μg N-(2-naphthyl)-hydroxylamine in arachis oil either once on the first day of life or once daily on the first 5 days of life resulted in an increase in the incidence of lung adenomas after 40 weeks (26% and

32%, respectively). When N-(2-naphthyl)-hydroxylamine in 3% aqueous gelatine was injected once daily for 5 days the incidence of lung adenomas was even higher (53%). The incidences observed in control groups given arachis oil or aqueous gelatine alone were 12.5% and 8.3%, respectively (Walters et al., 1967).

Intravesicular instillation of 5 mg N-(2-naphthyl)-hydroxylamine in dimethylsulphoxide every second week for 30 months resulted in the occurrence of invasive papillary transitional cell carcinomas in 3/4 treated dogs in 45 months. Four dogs treated likewise with the solvent alone did not develop bladder lesions (Radomski et al., 1971).

The evidence for a carcinogenic effect of several 2-amino-1-naphthol derivatives (i.e., hydrochloride, glucosiduronide, bis-(2-amino-1-naphthyl) phosphate, but not the 2-amino-1-naphthyl sulphate or 2-amino-1-naphthyl-dihydrogen phosphate) is based on the results obtained following their implantation into the bladder. The sulphate and phosphate conjugates were found to be non-carcinogenic (Bonser et al., 1963). The validity of this experimental model has been questioned, and these results are not unequivocal.

3.3 Observations in man[1]

(a) Case reports

There are many case reports linking 2-naphthylamine with bladder cancer in workmen manufacturing or using this chemical. Earlier reports were summarized by Hueper (1942). More recent cases are mentioned by Billiard-Duchesne (1960), Gehrmann et al.(1949), Tsuji (1963), Temkin (1963), Vigliani & Barsotti (1961), Veys (1969) and Scott (1962).

(b) Epidemiology

Case et al. (1954) showed that 2-naphthylamine induced bladder tumours in makers, users and purifiers of the chemical. Overall,

---

[1] See also preamble, "Carcinogenicity of the aromatic amines in man", of this volume.

there were 26 certified deaths from bladder cancer among 2-naphthyl-amine workers, where only 0.3 would have been expected from the whole male population of England and Wales (P <0.001). The morbidity was 55 cases. Mancuso & El-Attar (1967) in a further cohort study of a factory population reported that 2-naphthylamine had an "attack rate" of 952 per 100,000, which was higher than for benzidine. Goldwater et al. (1965) in a retrospective survey of a coal-tar dye plant showed that 12 out of 48 workmen exposed to 2-naphthylamine alone and not to other aromatic amines developed bladder tumours.

4. Comments on Data Reported and Evaluation

4.1 Animal data

2-Naphthylamine is carcinogenic in the mouse, hamster, dog and monkey. Given orally it has produced bladder carcinomas in the dog and monkey, and, at high dosage levels, in the hamster. By this route, it has increased the incidence of hepatomas in the mouse; in the rat and rabbit, it has little, if any, carcinogenic effect.

4.2 Human data

Epidemiological studies have shown that occupational exposure to 2-naphthylamine, either alone or when present as an impurity in other compounds, is strongly associated with the occurrence of bladder cancer. There is no doubt that 2-naphthylamine is a human bladder carcinogen.

5. References

Allen, M.J., Boyland, E., Dukes, C.E., Horning, E.S. & Watson, J.G. (1957) Cancer of the urinary bladder induced in mice with meta-bolites of aromatic amines and tryptophan. Brit. J. Cancer, 11, 212

Battye, R. (1966) Bladder carcinogens occurring during the production of "town" gas by coal carbonisation. In: Transcripts of the XV International Congress on Occupational Health, Vienna, 1966, Vol. VI-2, p. 153

Billiard-Duchesne, J.L. (1960) Cas français de tumeurs professionnelles de la vessie. Acta Un. int. Cancr., 16, 284

Bonser, G.M. (1943) Epithelial tumours of the bladder in dogs induced by pure β-naphthylamine. J. Path. Bact., 55, 1

Bonser, G.M., Boyland, E., Busby, E.R., Clayson, D.B., Grover, P.L. & Jull, J.W. (1963) A further study of bladder implantation in the mouse as a means of detecting carcinogenic activity: use of crushed paraffin wax or stearic acid as the vehicle. Brit. J. Cancer, 17, 127

Bonser, G.M., Clayson, D.B. & Jull, J.W. (1951) An experimental inquiry into the cause of industrial bladder cancer. Lancet, ii, 286

Bonser, G.M., Clayson, D.B., Jull, J.W. & Pyrah, L.N. (1952) The carcinogenic properties of 2-amino-1-naphthol hydrochloride and its parent amine 2-naphthylamine. Brit. J. Cancer, 6, 412

Bonser, G.M., Clayson, D.B., Jull, J.W. & Pyrah, L.N. (1956) The carcinogenic activity of 2-naphthylamine. Brit. J. Cancer, 10, 533

Boyland, E. (1958) The biochemistry of cancer of the bladder. Brit. med. Bull., 14, 153

Boyland, E., Dukes, C.E. & Grover, P.L. (1963) Carcinogenicity of 2-naphthylhydroxylamine and 2-naphthylamine. Brit. J. Cancer, 17, 79

Boyland, E., Kinder, C.H. & Manson, D. (1961) Synthesis and detection of di(2-amino-1-naphthyl)hydrogen phosphate, a metabolite of 2-naphthylamine in the dog. Biochem. J., 78, 175

Boyland, E. & Manson, D. (1966) The metabolism of 2-naphthylamine and 2-naphthylhydroxylamine derivatives. Biochem. J., 101, 84

Butt, L.T. & Strafford, N. (1956) Papilloma of the bladder in the chemical industry. Analytical methods for the determination of benzidine and β-naphthylamine, recommended by the ABCM Sub-Committee. J. appl. Chem., 6, 525

Case, R.A.M., Hosker, M.E., McDonald, D.B. & Pearson, J.T. (1954) Tumours of the urinary bladder in workmen engaged in the manufacture and use of certain dyestuff intermediates in the British chemical industry. I. The role of aniline, benzidine, alpha-naphthylamine and beta-naphthylamine. Brit. J. industr. Med., 11, 75

Conzelman, G.M., Jr & Moulton, J.E. (1972) Dose-response relationships of the bladder tumorigen 2-naphthylamine: a study in beagle dogs. J. nat. Cancer Inst., 49, 193

Conzelman, G.M., Jr, Moulton, J.E., Flanders, L.E., III, Springer, K. & Crout, D.W. (1969) Induction of transitional cell carcinomas of the urinary bladder in monkeys fed 2-naphthylamine. J. nat. Cancer Inst., 42, 825

Deichmann, W.B. & Radomski, J.L. (1969) Carcinogenicity and metabolism of aromatic amines in the dog. J. nat. Cancer Inst., 43, 263

Gehrmann, G.H., Foulger, J.H. & Fleming, A.J. (1949) Occupational carcinoma of the bladder. In: Proceedings of the Ninth International Congress on Industrial Medicine, London, 1948, Bristol, Wright, p. 427

Ghetti, G., Bartalini, E., Armeli, G. & Pozzoli, L. (1968) Separazione e dosaggio in vari substrati di amine aromatiche (benzidina, o-tolidina, dianisidina, diclorobenzidina, α-naftilamina, β-naftilamina). Messa a punto di nuovi metodi analitici per l'igiene del lavoro. Lav. umano, 20, 389

Goldwater, L.J., Rosso, A.J. & Kleinfeld, M. (1965) Bladder tumors in a coal-tar dye plant. Arch. environm. Hlth, 11, 814

Gupta, R.C. & Srivastava, S.P. (1971) Oxidation of aromatic amines by peroxodisulphate ion. Z. analyt. Chem., 257, 275

Hadidian, Z., Fredrickson, T.N., Weisburger, E.K., Weisburger, J.H., Glass, R.M. & Mantel, N. (1968) Tests for chemical carcinogens. Report on the activity of derivatives of aromatic amines, nitrosamines, quinolines, nitroalkanes, amides, epoxides, aziridines and purine antimetabolites. J. nat. Cancer Inst., 41, 985

Hoffmann, D., Masuda, Y. & Wynder, E.L. (1969) α-Naphthylamine and β-naphthylamine in cigarette smoke. Nature (Lond.), 221, 254

Hueper, W.C. (1942) Occupational Tumours and Allied Diseases, Springfield, Illinois, Thomas

Hueper, W.C., Wiley, F.H. & Wolfe, H.D. (1938) Experimental production of bladder tumors in dogs by administration of beta-naphthylamine. J. industr. Hyg., 20, 46

Hyatt, J.C. (1973) US mulls rules for handling of chemicals that can lead to cancer in plant workers. Wall Street J., April 11, 36

Kilner, E. & Samuel, D.M. (1960) Applied Organic Chemistry, New York, Interscience, pp. 262 and 266

Mancuso, T.F. & El-Attar, A.A. (1967) Cohort study of workers exposed to beta-naphthylamine and benzidine. J. occup. Med., 9, 277

Masuda, Y. & Hoffmann, D. (1969) Quantitative determination of 1-naphthylamine and 2-naphthylamine in cigarette smoke. Analyt. Chem., 41, 650

Masuda, Y., Mori, K. & Kuratsune, M. (1967) Studies on bladder carcinogens in the human environment. I. Naphthylamines produced by pyrolysis of amino acids. Int. J. Cancer, 2, 489

Radomski, J.L. & Brill, E. (1971) The role of N-oxidation products of aromatic amines in the induction of bladder cancer in the dog. Arch. Toxicol., 28, 159

Radomski, J.L., Brill, E., Deichmann, W.B. & Glass, E.M. (1971) Carcinogenicity testing of N-hydroxy and other oxidation and decomposition products of 1- and 2-naphthylamine. Cancer Res., 31, 1461

Radomski, J.L., Rey, A.A. & Brill, E. (1973) Conjugates of the N-hydroxy metabolites of aromatic amines in dog urine. Proc. amer. Ass. Cancer Res., 14, 126

Saffiotti, U., Cefis, F., Montesano, R. & Sellakumar, A.R. (1967) Induction of bladder cancer in hamsters fed aromatic amines. In: Deichmann, W. & Lampe, K.F., eds., Bladder Cancer. A Symposium, Birmingham, Alabama, Aesculapius, p.129

Sawicki, E., Johnson, H. & Kosinski, K. (1966) Chromatographic separation and spectral analysis of polynuclear aromatic amines and heterocyclic imines. Microchem. J., 10, 72

Scott, T.S. (1962) Carcinogenic and chronic toxic hazards of aromatic amines, Amsterdam, New York, Elsevier, p. 61

Sellakumar, A.R., Montesano, R. & Saffiotti, U. (1969) Aromatic amines carcinogenicity in hamsters. Proc. amer. Ass. Cancer Res., 10, 78

Shimomura, K. & Walton, H.F. (1968) Thin-layer chromatography of amines by ligand exchange. Separation Science, 3, 493

The Society of Dyers and Colourists (1971) Colour Index, 3rd ed., 4, 4807

Takemura, N., Akiyama, T. & Nakajima, C. (1965) A survey of the pollution of the Sumida river, especially on the aromatic amines in the water. Int. J. air wat. Poll., 9, 665

Temkin, I.S. (1963) Industrial Bladder Carcinogenesis, Oxford, London, New York, Paris, Pergamon

Treibl, H.G. (1967) Naphthalene derivatives. In: Kirk, R.E. & Othmer, D.F., eds., Encyclopedia of Chemical Technology, 2nd ed., New York, John Wiley & Sons, Vol. 13, p. 708

Troll, W. & Nelson, N. (1961) N-hydroxy-2-naphthylamine, a urinary metabolite of 2-naphthylamine in man and dog. Fed. Proc., 20, 41

Troll, W., Tessler, A.N. & Nelson, N. (1963) Bis(2-amino-1-naphthyl)phosphate, a metabolite of beta-naphthylamine in human urine. J. Urol., 89, 626

Tsuji, I. (1963) Environmental and industrial cancer of the bladder in Japan. Acta Un. int. Cancr., 18, 662

UICC (1970) The quantification of environmental carcinogens (UICC Technical Report Series, Vol. 4)

US Government (1973) Occupational safety and health standards. US Federal Register, 38, No.85, 10929

US Tariff Commission (1922) Census of Dyes and other Synthetic Organic Chemicals, 1921, Tariff Information Series No.26

US Tariff Commission (1956) Synthetic Organic Chemicals, United States Production and Sales, 1955, Report No. 198, Second Series

US Tariff Commission (September 1968) Imports of Benzenoid Chemicals and Products, 1967, TC Publication 264

Veys, C.A. (1969) Two epidemiological inquiries into the incidence of bladder tumors in industrial workers. J. nat. Cancer Inst., 43, 219

Vigliani, E.D. & Barsotti, M. (1961) Environmental tumors of the bladder in some Italian dye-stuff factories. Acta Un. int. Cancr., 18, 669

Walters, M.A., Roe, F.J.C., Mitchley, B.C.V. & Walsh, A. (1967) Further tests for carcinogenesis using newborn mice: 2-naphthylamine, 2-naphthylhydroxylamine, 2-acetylaminofluorene and ethyl methane sulphonate. Brit. J. Cancer, 21, 367

4-NITROBIPHENYL [*]

1. Chemical and Physical Data

1.1 Synonyms and trade names

Chem. Abstr. No.: 92-93-3

p-Nitrobiphenyl; p-nitrodiphenyl; 4-nitrodiphenyl; PNB;
4-phenyl-nitrobenzene; p-phenyl-nitrobenzene

1.2 Chemical formula and molecular weight

 $C_{12}H_9NO_2$ Mol. wt: 199.2

1.3 Chemical and physical properties of the pure substance

(a) Description: White needles

(b) Melting-point: $113.8^\circ C$

(c) Boiling-point: $340^\circ C$

(d) Solubility: Almost insoluble in water; slightly soluble in cold alcohol; readily soluble in benzene, chloroform and ether

(e) Chemical reactivity: Can be reduced to 4-aminobiphenyl. Prolonged warming with powdered KOH in benzene leads to the formation of 3-hydroxy-4-nitrobiphenyl

1.4 Technical products and impurities

No information is available to the Working Group on technical products or impurities. It is possible that isomeric nitrobiphenyls may occur as impurities.

---

[*] Considered by the Working Group in Lyon, June 1973.

2. Production, Use, Occurrence and Analysis

(a) Production and use[1]

4-Nitrobiphenyl can be prepared by the nitration of diphenyl. No evidence has been found that 4-nitrobiphenyl is produced commercially in the United States at the present time. One US manufacturer of laboratory chemicals included it in his 1970 catalogue but had dropped it by 1972.

The only known commercial use for 4-nitrobiphenyl is as a chemical intermediate in the preparation of 4-aminobiphenyl (C_6H_5-$C_6H_4NH_2$), which was formerly used in large quantities, primarily as a rubber antioxidant. According to one source, the two manufacturers of 4-aminobiphenyl ceased production when investigators reported that it was a bladder carcinogen in the early 1950's (Hueper, 1969). Another source reported that 4-amino-biphenyl was produced in commercial quantities in the US in only a single plant which operated from 1935 to 1955 (the source of the raw material, 4-nitrobiphenyl, which was reportedly shipped in by tank car, was not identified). In the plant where the nitro compound was reduced to the amine, closed systems were used so that day-to-day exposures were as low as possible (Melick et al., 1955, 1971).

(b) Occurrence

In 1862, residues from the distillation of aniline were found to contain 4-aminobiphenyl (Hofmann, 1862). In 1952, it was suggested that the crude aniline produced commercially prior to 1900 may have contained 4-aminobiphenyl as a result of the reduction of 4-nitrobiphenyl produced by the nitration of impure benzene containing diphenyl. The absence of 4-aminobipheynl in residues from the distillation of commercial aniline in 1952 was taken to indicate that improved distillation techniques had eliminated the diphenyl from the benzene (Walpole et al., 1952).

---

[1] Data from Chemical Information Services, Stanford Research Institute, USA.

The presence, use and manufacture of 4-nitrobiphenyl in factories is prohibited in the UK under the Carcinogenic Substances Regulations 1967 - Statutory Instrument (1967), No. 487. It is also listed with 13 other compounds in the US Federal Register (US Government, 1973) as subject to an emergency temporary standard on certain carcinogens under an order made by the Occupational Safety and Health Administration, Department of Labor, on 26 April, 1973.

(c) Analysis

Aromatic nitro compounds may be quantitatively converted to aromatic amines and subsequently determined using guidelines suggested for the latter class of compounds (UICC, 1970). Improved techniques for the estimation of 4-nitrobiphenyl using absorption spectrometry have been developed (Uehleke & Nestel, 1967).

3. Biological Data Relevant to the Evaluation
of Carcinogenic Risk to Man

3.1 Carcinogenicity and related studies in animals

(a) Oral administration

Dog: Deichmann et al.(1958) induced carcinomas of the bladder epithelium in 2/4 female mongrel dogs in 33 months by feeding a dose of 300 mg 4-nitrobiphenyl by capsule 3 times per week for life (maximum total dose, approximately 10 g/kg bw). Subsequently, it was shown that a total dose of 0.77 g/kg bw was not effective in inducing these tumours in 6 dogs within 3 years (Deichmann et al., 1965).

3.2 Other relevant biological data

(a) Animals

Laham (1960) presented chromatographic evidence that in rats 4-nitrobiphenyl was converted to 4-aminobiphenyl and 4-aminobiphenyl-3-yl hydrogen sulphate. Uehleke & Nestel (1967) demonstrated that in vitro 25% of 4-nitrobiphenyl was reduced to 4-aminobiphenyl within 40 minutes during anaerobic reduction in the presence of rat liver enzymes and the necessary cofactors. The presence of 4-nitrosobiphenyl was demonstrated during the reduction. 4-Nitrosobiphenyl was, however, reduced four times faster than was the 4-nitro compound. Small amounts of non-acetylated N-hydroxy products have been identified in the urine of dogs and monkeys given a single dose of 5 mg/ kg bw 4-nitrobiphenyl (Radomski et al., 1973).

(b) Carcinogenicity of metabolites

The carcinogenicity of 4-aminobiphenyl and its metabolites has been reviewed previously (IARC, 1972).

3.3 Observations in man

There are no reports known to the Committee on carcinogenicity of 4-nitrobiphenyl in man. Because of the conversion of 4-nitrobiphenyl to 4-aminobiphenyl it is not possible to separate the exposures to either substance. Epidemiological studies in workers exposed to 4-aminobiphenyl have been reviewed previously (IARC, 1972).

4. Comments on Data Reported and Evaluation[1]

4.1 Animal data

4-Nitrobiphenyl induced carcinomas of the bladder when given orally to dogs, the only species and route known to have been tested.

4.2 Human data

There are no data on the carcinogenicity of 4-nitrobiphenyl in man. However, it has been used in the production of 4-aminobiphenyl, which is a recognized human bladder carcinogen (IARC, 1972).

---

[1] See also the section "Extrapolation from animals to man" in the introduction to this volume.

5. References

Deichmann, W.B., MacDonald, W.M., Coplan, M.M., Woods, F.M. & Anderson, W.A.D. (1958) Para nitrobiphenyl, a new bladder carcinogen in the dog. Industr. Med. Surg., 27, 634

Deichmann, W.B., Radomski, J., Glass, E., Anderson, W.A.D., Coplan, M.M. & Woods, F.M. (1965) Synergism among oral carcinogens. III. Simultaneous feeding of four bladder carcinogens to dogs. Industr. Med. Surg., 34, 640

Hofmann, A.W. (1862) De quelques produits secondaires formés dans la fabrication de l'aniline. C.R. Acad. Sci.(Paris), 55, 901

Hueper, W.C. (1969) Occupational and environmental cancers of the urinary system, New Haven, London, Yale University Press

International Agency for Research on Cancer (1972) IARC Monographs on the Evaluation of Carcinogenic Risk of Chemicals to Man, 1, p. 74

Laham, S. (1960) Biological conversion of 4-nitrobiphenyl to an active carcinogen. Canad. J. Biochem., 38, 1383

Melick, W.F., Escue, H.M., Naryka, J.J., Mezera, R.A. & Wheeler, E.P. (1955) The first reported cases of human bladder tumors due to a new carcinogen - xenylamine. J. Urol., 74, 760

Melick, W.F., Naryka, J.J. & Kelley, R.E. (1971) Bladder cancer due to exposure to para-aminobiphenyl: a 17-year follow up. J. Urol., 106, 220

Radomski, J.L., Conzelman, G.M., Jr, Rey, A.A. & Brill, E. (1973) N-Oxidation of certain aromatic amines, acetamides and nitro compounds by monkeys and dogs. J. nat. Cancer Inst., 50, 989

Uehleke, H. & Nestel, K. (1967) Hydroxylamino- und Nitrosobiphenyl: Biologische Oxydationsprodukte von 4-Aminobiphenyl und Zwischenprodukte der Reduktion von 4-Nitrobiphenyl. Naunyn-Schmiedebergis Arch. Pharmak. exp. Path., 257, 151

UICC (1970) The quantification of environmental carcinogens (UICC Technical Report Series, Vol. 4)

US Government (1973) Occupational safety and health standards. US Federal Register, 38, No. 85, 10929

Walpole, A.L., Williams, M.H.C. & Roberts, D.C. (1952) The carcinogenic action of 4-aminodiphenyl and 3:2'-dimethyl-4-aminodiphenyl. Brit. J. industr. Med., 9, 255

N,N-BIS(2-CHLOROETHYL)-2-NAPHTHYLAMINE[*]

1. Chemical and Physical Data

1.1 Synonyms and trade names

Chem. Abstr. No.: 49-40-31

CB 1048; chlornaftina; chlornaphazin; chlornaphazine; chlor-naphthin; chloronaftina; chloronaphthine; dichloroethyl-beta-naphthylamine; di(2-chloroethyl)-beta-naphthylamine; naphthylamine mustard; beta-naphthyl-bis(beta-chloroethyl)amine; beta-naphthyl-di-(2-chloroethyl)amine

Erysan; R48

1.2 Chemical formula and molecular weight

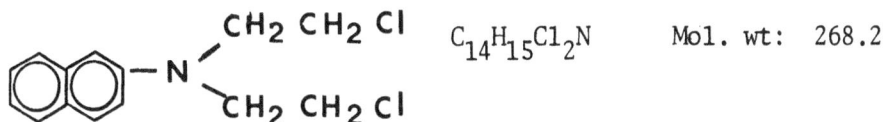

$C_{14}H_{15}Cl_2N$ Mol. wt: 268.2

1.3 Chemical and physical properties of the pure substance

(a) Description: Colourless plates

(b) Boiling-point: $210^{\circ}C$

(c) Melting-point: $54-56^{\circ}C$

(d) Solubility: Very sparingly soluble in water and glycerol; more soluble in petroleum ether, ethanol, olive oil, ether, acetone and benzene (in ascending order)

---

[*] Considered by the Working Group in Lyon, June 1973.

(e) Chemical reactivity: A nitrogen mustard; it reacts with nucleophiles. A detailed study of the hydrolysis in aqueous acetone has been made (Ross, 1949).

1.4 Technical products and impurities

No information is available to the Working Group.

2. Production, Use, Occurrence and Analysis

(a) Production and use[1]

No evidence was found that it has ever been produced commercially or found commercial usage in the United States. However, this compound has been used in a number of countries as a chemotherapeutic agent for the treatment of leukaemia and of Hodgkin's disease (Videbaek, 1964) as well as for the control of polycythaemia vera (Thiede et al., 1964). It has reportedly been withdrawn from use in Denmark because of detected instances of cancer of the bladder following its administration. Although it still appears in some drug directories, N,N-bis(2-chloroethyl)-2-naphthyl-amine probably does not have wide therapeutic usage.

(b) Occurrence

N,N-bis(2-chloroethyl)-2-naphthylamine has not been reported to occur as such in nature.

(c) Analysis

No methods specific for the analysis of chlornaphazine are known to the Working Group. General methods for the detection of alkylating agents are available (Preussmann et al., 1969; Sawicki & Sawicki, 1969).

---

[1] Data from Chemical Information Services, Stanford Research Institute, USA.

3. Biological Data Relevant to the Evaluation
of Carcinogenic Risk to Man

3.1 Carcinogenicity and related studies in animals

(a) Subcutaneous and/or intramuscular injection

Rat: Koller (1953) noted that repeated s.c. injections of 40 mg N,N-bis(2-chloroethyl)-2-naphthylamine induced local subcutaneous sarcomas in 7 rats (no other details are reported).

(b) Intraperitoneal administration

Mouse: In 4 groups of 30 A/J mice given total doses of 280, 1119, 4477 and 17,900 μM/kg bw (equivalent to 75, 300, 1200 and 4800 mg/kg bw) in aqueous solution by i.p. injection 3 times weekly for 4 weeks, a dose related increase in the incidence of pulmonary adenomas, and in the number of lung tumours per mouse, was observed. The mice were first injected when 4 to 6 weeks old and were killed at 39 weeks after the first injection. The respective incidences of lung tumours per mouse in the treated groups were 0.5, 0.9, 2.0 and 3.6, and 0.48 in the vehicle control group (Shimkin et al., 1966).

3.2 Other relevant biological data

(a) Animals

Rats injected with N,N-bis(2-chloroethyl)-2-naphthylamine excreted 2-amino-1-naphthyl hydrogen sulphate and 2-acetamido-6-naphthyl hydrogen sulphate (Boyland & Manson, 1963).

3.3 Observations in man

(a) Case reports

Eleven separate cases of bladder cancer associated with the use of chlornaphazine were reported by Thiede et al.(1964), Videbaek (1964) and Laursen (1970). Three further cases have been identified by Thiede & Christensen (1969).

(b) Epidemiology

Thiede et al.(1964) collected data on 61 polycythaemia patients
treated with chlornaphazine and showed that there was no association
with cigarette smoking or occupation in the 7 who developed bladder
tumours. In 40 cases of polycythaemia not treated with chlornaphazine there were no cases of bladder tumours. These patients all
survived the onset of the disease by at least 10 years. In 1964, the
dose necessary to induce bladder tumours was thought to be more than
100 g, but a recent report (Thiede & Christensen, 1969) described a
further 3 cases of bladder tumours among the original 61 patients
and 5 further patients with abnormal urinary cytology. One tumour
case and 1 patient with abnormal urinary cytology had received only
4 and 2 g of the drug, respectively.

These studies (Thiede et al., 1964; Thiede & Christensen, 1969)
can be summarized as follows:-

Number of patients	61
Number surviving in 1969	27
Cases of bladder tumour	10
Additional cases with abnormal urinary cytology	5
Latent period from adminis-tration of drug to diagnosis of tumour	5.5 (range 3-10) years
Dose of drug	2-350 g

A common factor among the 10 patients in whom bladder tumours
have been diagnosed is that 9 of the patients were also treated with
^{32}P as sodium phosphate injections. However, Thiede et al.(1964)
report that out of 46 patients treated with ^{32}P as sodium phosphate
alone without chlornaphazine, no cases of bladder tumours were
recorded.

122

4. Comments on Data Reported and Evaluation

4.1 Animal data

N,N-Bis(2-chloroethyl)-2-naphthylamine is carcinogenic to the mouse lung by the intraperitoneal route and has a local carcinogenic effect in rats following subcutaneous injection.

4.2 Human data

This drug has been administered to man with ^{32}P sodium phosphate for the treatment of polycythaemia and of neoplasias of the haemopoietic system in doses of up to 400 mg/day. Follow-up studies have shown that under these conditions, the drug is carcinogenic, producing bladder tumours after administration of total doses as low as 4 g.

5. References

Boyland, E. & Manson, D. (1963) Metabolism of 2-naphthylamine and its derivatives. A.R. Brit. Emp. Cancer Campgn, 41, 69

Koller, P.C. (1953) Dicentric chromosomes in a rat tumour induced by an aromatic nitrogen mustard. Heredity, 6, 181

Laursen, B. (1970) Cancer of the bladder in patients treated with chlornaphazine. Brit. med. J., iii, 684

Preussmann, R., Schneider, H. & Epple, F. (1969) Untersuchungen zum Nachweis alkylierender Agentien. II. Der Nachweis verschiedener Klassen alkylierender Agentien mit einer Modifikation der Farbreaktion mit 4-(4-Nitrobenzyl)-pyridin (NBP). Arzneimittel-Forsch., 19, 1059

Ross, W.C.J. (1949) Aryl-2-halogenoalkylamines. I. J. chem. Soc.(Lond.), 43, 183

Sawicki, E. & Sawicki, C.R. (1969) Analysis of alkylating agents. Application to air pollution. Ann. N.Y. Acad. Sci., 163, 895

Shimkin, M.B., Weisburger, J.H., Weisburger, E.K., Gubareff, N. & Suntzeff, V. (1966) Bioassay of 29 alkylating chemicals by the pulmonary-tumour response in strain A mice. J. nat. Cancer Inst., 36, 915

Thiede, T., Chievitz, E. & Christensen, B.C. (1964) Chlornaphazin as a bladder carcinogen. Acta med. scand., 175, 721

Thiede, T. & Christensen, B.C. (1969) Bladder tumours induced by chlor-
    naphazin: a five-year follow-up study of chlornaphazin-treated
    patients with polycythaemia. Acta med. scand., 185, 133

Videbaek, A. (1964) Chlornaphazin (Erysan [R]) may induce cancer of the
    urinary bladder. Acta med. scand., 176, 45

HYDRAZINE

AND

DERIVATIVES

HYDRAZINE[*]

1. Chemical and Physical Data

1.1 Synonyms and trade names

Chem. Abstr. No.: 302-01-2

Diamide; diamine; hydrazine base

1.2 Chemical formula and molecular weight

H_2N-NH_2 N_2H_4 Mol. wt: 32.0

1.3 Chemical and physical properties of the pure substance

(a) Description: A clear, colourless, hygroscopic liquid with the characteristic ammonia-like, fishy odour of alkyl hydrazines; fumes in air

(b) Boiling-point: $113.5^\circ C$

(c) Melting-point: $2^\circ C$

(d) Density: d_4^{20} 1.0083, d_4^{15} 1.011

(e) Refractive index: n_D^{35} 1.4644

(f) Solubility: Miscible with water and ethanol; slightly miscible with hydrocarbons and halogenated hydrocarbons

(g) Stability: If air is excluded it can be stored, preferably in the form of the hydrate, without decomposition for a long time in paraffin bottles. Reacts exothermically and violently with metal oxides (e.g., iron oxides, molybdenum oxides) and oxidizing agents

(h) Volatility: Vapour pressure is 10.4 mm Hg at $20^\circ C$

---

[*] Considered by the Working Group in Lyon, June 1973.

(i) Chemical reactivity: The compound is a powerful reducing agent. It reacts with acids to yield salts, for example, hydrazine sulphate. A survey of the physical and chemical properties of hydrazine has been published (Harshman, 1957).

1.4 Technical products and impurities

Hydrazine is produced in the United States in a propellant grade which contains a minimum of 97.5% of the active ingredient. The major impurity is water (up to 2.5%), and up to 0.2% of undefined insoluble material may also be present.

Aqueous solutions containing variable quantities of hydrazine are also offered for sale for a variety of industrial uses. Such products are usually identified by their content of hydrazine hydrate, $H_2NNH_2.H_2O$ (which corresponds to a 62.7% hydrazine content). One manufacturer's data sheet indicates that such aqueous solutions contain a maximum of 0.01% nonvolatile matter, a maximum of 0.001% chloride ion, and 0.005%, maximum, of each of the following: sulphate ion, iron, and heavy metals (as lead).

2. Production, Use, Occurrence and Analysis

Comprehensive reviews on the manufacture, chemistry and uses of hydrazine have been published (Reed, 1957; Raphaelian, 1966).

(a) Production and use[1]

Hydrazine was discovered in 1887, and a synthetic method was developed in 1907, but it did not become a significant commercial chemical until World War II when interest in its use as a rocket fuel prompted large scale production and markedly lower prices.

Although one small US producer is believed to produce hydrazine by the oxidation of urea with hypochlorite (a process used earlier by some other companies), most of the hydrazine produced in the US is made by the

---

[1] Data from Chemical Information Services, Stanford Research Institute, USA.

Raschig process. In this process, ammonia is first converted to chloramine by treatment with sodium hypochlorite, and then further reaction of
this chloramine with excess ammonia and sodium hydroxide produces an
aqueous solution of hydrazine and by-product sodium chloride. Fractional
distillation of the reaction product produces hydrazine hydrate solutions.
Azeotropic distillations (e.g., with aniline), or other methods, are used
to remove the water and produce propellant grades of hydrazine. Variations
of the Raschig process involving the formation of intermediate diazacyclopropanes are in use in other countries.

In 1966, it was estimated that total annual production capacity of
the five US producers was 10 million kg of hydrazine (100% basis), and
that 7 million kg per year of this was made in one plant built specifically
to manufacture hydrazine for use in rockets (Raphaelian, 1966). Since
that time the demand for hydrazine for rocket fuel has dropped dramatically,
and the plant producing 7 million kg per year is believed to have been
shut down. Some of the smaller plants may also have been shut down, since
one source indicated that only two US companies were offering hydrazine
hydrate for sale in 1971, and that the product from these companies was
having to compete with material imported from Japan (Anon., 1971).

In 1966, it was reported (Raphaelian, 1966) that hydrazine was being
produced in the following countries (the number of producing companies
and total annual capacity based on 100% hydrazine are shown in parentheses):
Japan (3 producers, 2 million kg); the Federal Republic of Germany (1
producer, 2 million kg); United Kingdom (1 producer, 2 million kg); France
(2 producers, 0.5 million kg); and Spain (1 producer, 0.04 million kg).
In 1970, it was reported that almost half of the hydrazine produced in the
United Kingdom was being exported, and United Kingdom consumption was
estimated at 1 million kg per year (Anon., 1970).

In 1966, US consumption of hydrazine (100% basis) was estimated
(Raphaelian, 1966) to have been 7 million kg per year with the following
consumption pattern (millions of kg consumed shown in parentheses):
rocket fuel (5); manufacture of maleic hydrazide (0.5); manufacture of
azobisformamide blowing agent (0.4); boiler water treatment (0.2); manu-

facture of 3-amino-1,2,4-triazole herbicide (0.2); manufacture of other blowing agents (0.2); manufacture of medicinals, e.g., isoniazid anti-tubercular agent (0.2); and the remainder in other uses. The usage in rocket fuels has since dropped to a level estimated to be less than 0.5 million kg per year, and the other markets do not appear to have grown very rapidly. In 1971, one source estimated the US market for hydrazine hydrate to be only 2.3 million kg (corresponding to about 1.4 million kg of hydrazine). This same source estimated that 40% of the hydrazine hydrate was being used as an oxygen scavenger in boiler feedwater treatment and 60% was being used as a chemical intermediate, e.g., as a reducing agent, and in the production of the hydrazine salts used in soldering fluxes (Anon., 1971).

It has been stated that hydrazine's greatest industrial potential lies in its use as a fuel for fuel cells, in which high current densities result when hydrazine is catalytically decomposed and oxidized into nitrogen and water. This potential is unlikely to develop, however, unless the price of hydrazine becomes considerably lower than it has been in the past.

(b) Occurrence

Hydrazine has been found to be a primary product of nitrogen fixation by Azotobacter agile. The use of hydrazine in boiler water treatment might result in its brief appearance in waste discharge, but it would react with oxygen rapidly. The use of hydrazine as a chemical intermediate would not be likely to result in its appearance in unreacted form in the environment. One source has reported that the burning of rocket fuels based on hydrazine and dimethylhydrazine produces exhaust gases which contain only trace quantities of unchanged fuel (Anon., 1969).

(c) Analysis

Hydrazine and hydrazine sulphate can be determined by titration, colorimetry or potentiometry (Feinsilver et al., 1959; Nair & Nair, 1971). Small amounts (below 1 ppm) can be estimated photometrically after reaction with p-dimethylaminobenzaldehyde to yield an intense red colour with a minimum transmittance at 445 nm (Wood, 1953) or 455 nm (Spinkova, 1971). Hydrazine can be quantitatively estimated using gas chromatography (Bighi & Saglietto, 1966).

3. Biological Data Relevant to the Evaluation
of Carcinogenic Risk to Man

3.1 Carcinogenicity and related studies in animals

(a) Oral administration

Mouse: Adequate studies in different strains of mice have demonstrated that hydrazine given mainly as the hydrazine sulphate (HS) produces a high incidence of multiple pulmonary adenomas and adenocarcinomas (Biancifiori & Severi, 1966). Lung tumours have been induced in Swiss mice (Roe et al., 1967; Toth, 1969, 1971), in CBA/Cb/Se mice (Biancifiori et al., 1963a; 1964; Biancifiori, 1969), in BALB/c/Cb/Se mice (Biancifiori, 1970a,b,c; Biancifiori et al., 1963b; Biancifiori & Ribacchi, 1962a,b), in C3Hb/Cb/se mice (Biancifiori, 1971), in (BALB/c x DBA/2)F_1 hybrids (Kelly et al., 1969) and in A/J mice (Yamamoto & Weisburger, 1970).

In groups of 19 to 26 male and female BALB/c/Cb/Se mice, the incidence of pulmonary tumours produced by HS was shown to be dose dependent. The lowest dose used (21 mg) produced an incidence of 54% in males and 32% in females surviving up to 78 weeks. In control mice 24% of the males and 4% of the females developed pulmonary tumours, mostly between 90 and 100 weeks (Biancifiori, 1970a).

Hepatomas and hepatocarcinomas have been observed in 3 strains of mice treated orally with HS. These include BALB/c/Cb/Se mice (Biancifiori, 1970a,b), CBA/Cb/Se mice (Biancifiori, 1970d; Biancifiori et al., 1964; Severi & Biancifiori, 1967, 1968) and C3Hb/Cb/Se mice (Biancifiori, 1971).

In a group of 40 female C3H mice, 0.012% HS given in the drinking-water for life greatly increased the incidence of lung tumours but reduced the incidence of mammary tumours from 77% in the controls to 38% in the treated mice (Toth, 1969). HS does not show initiating activity when administered orally to BALB/c/Cb/Se mice and followed by skin application of croton oil (Biancifiori et al., 1964).

131

Newborn mice: Milia (1965) treated 25 newborn BALB/c/Cb/Se mice with increasing doses of HS ranging from 25-600 µg/day by stomach tube for 60 days. A total dose of 17 mg was administered. The mice were killed at 18 weeks of age, and 96% had lung tumours with an average of 3 tumours per mouse.

Rat: Severi & Biancifiori (1968) administered daily doses of 18 mg or 12 mg HS by stomach tube to 14 male and 18 female Cb/Se rats for 68 weeks. Lung tumours (adenomas and adenocarcinomas) were observed in 3/14 males and in 5/18 females in 109 weeks. Out of 13 male and 13 female rats, hepatic cell carcinomas or spindle cell sarcomas were observed in 4 male rats. No lung or liver tumours were found in untreated controls (28 males and 22 females) surviving up to 104 weeks.

Hamster: Biancifiori (1970d) and Toth (1972) administered HS orally to Syrian golden hamsters and noted no significant increase in the number of tumours produced in either experiment.

(b)   Intraperitoneal administration

Mouse: Juhász et al.(1966) and Juhász (1967) injected 30 male and 30 female white mice with 0.5 mg hydrazine in physiological saline. A total dose of 400 mg/kg bw was given in 16 separate doses over 46 days. Of 13/34 survivors, 4 mice developed reticulum cell sarcomas of the mediastinum and 9 mice developed myeloid leukaemias within 100 to 313 days. A thymic lymphoma was observed in 1/60 control mice. An increase in the number of lung tumours was also observed in other strains of mice, namely (BALB/c x DBA/2) F_1 hybrid, C57BL, SWR and BALB/c /Cb/Se (newborn) (Kelly et al., 1969; Mirvish et al., 1969; Milia et al., 1965).

3.2   Other relevant biological data

(a)   Animals

Absorption of hydrazine through the skin in dogs is rapid, and the hydrazine can be detected in femoral blood within 30 seconds (Smith & Clark, 1972).

Hydrazine is possibly degraded to ammonia, as evidenced by the elevation of blood ammonia in dogs given hydrazine; however, diacetyl-hydrazine is not (McKennis et al., 1959; Colvin, 1969).

3.3 Observations in man

No data are available to the Working Group.

4. Comments on Data Reported and Evaluation[1]

4.1 Animal data

Hydrazine or hydrazine salts have been shown to be carcinogenic in mice after oral and intraperitoneal administration, and in rats following oral administration. No tumours were observed in Syrian golden hamsters after oral administration.

4.2 Human data

No epidemiological data are available to the Working Group.

---

[1] See also the section "Extrapolation from animals to man" in the introduction to this volume.

5. References

Anon. (1969) Chemical and Engineering News, July 14, 16

Anon. (1970) European Chemical News, June 12, 6

Anon. (1971) Oil, Paint and Drug Reporter, September 6, 15

Biancifiori, C. (1969) Esistenza di un fattore ormonico nella cancero-genesi polmonare da idrazina. Lav. Ist. Anat. Univ. Perugia, 29, 29

Biancifiori, C. (1970a) Tumori polmonari ed epatici da idrazina solfato a dosi ridotte in topi BALB/c/Cb/Se. Lav. Ist. Anat. Univ. Perugia, 30, 89

Biancifiori, C. (1970b) Ovarian influence on pulmonary carcinogenesis by hydrazine sulfate in BALB/c/Cb/Se mice. J. nat. Cancer Inst., 45, 965

Biancifiori, C. (1970c) Cancerogenesi da idrizina solfato in trapianti isogenici tracheo-broncopolmonari in topi BALB/c/Cb/Se. Lav. Ist. Anat. Univ. Perugia, 30, 137

Biancifiori, C. (1970d) Hepatomas in CBA/Cb/Se mice and liver lesions in golden hamsters induced by hydrazine sulfate. J. nat. Cancer Inst., 44, 943

Biancifiori, C. (1971) Influenza degli ormoni ovarici nella cancerogenesi polmonare da idrazina solfato in topi C3Hb/Cb/Se. Lav. Ist. Anat. Univ. Perugia, 31, 5

Biancifiori, C., Bucciarelli, E., Clayson, D.B. & Santilli, F.E. (1964) Induction of hepatomas in CBA/Cb/Se mice by hydrazine sulphate and the lack of effect of croton oil on tumour induction in BALB/c/Cb/Se mice. Brit. J. Cancer, 18, 543

Biancifiori, C., Bucciarelli, E., Santilli, F.E. & Ribacchi, R. (1963a) Carcinogenesi polmonare da idrazide dell' acido isonicotinico (INI) e suoi metaboliti in topi CBA/Cb/Se substrain. Lav. Ist. Anat. Univ. Perugia, 23, 209

Biancifiori, C. & Ribacchi, R. (1962a) The induction of pulmonary tumours in BALB/c mice by oral administration of isoniazid. In: Severi, L., ed., The Morphological Precursors of Cancer, Perugia, Division of Cancer Research, p. 635

Biancifiori, C. & Ribacchi, R. (1962b) Pulmonary tumours in mice induced by oral isoniazid and its metabolites. Nature(Lond.), 194, 488

Biancifiori, C., Ribacchi, R., Bucciarelli, E., DiLeo, F.P. & Milia, U. (1963b) Cancerogenesi polmonare da idrazina solfato in topi femmine BALB/c. Lav. Ist. Anat. Univ. Perugia, 23, 115

Biancifiori, C. & Severi, L. (1966) The relation of isoniazid (INH) and allied compounds to carcinogenesis in some species of small laboratory animals: A review. Brit. J. Cancer, 20, 528

Bighi, C. & Saglietto, G. (1966) The effect of alkalized adsorbent on the gas-chromatographic separation of hydrazine derivatives. J. gas Chromat., 4, 303

Colvin, L.B. (1969) Metabolic fate of hydrazines and hydrazides. J. pharm. Sci., 58, 1433

Feinsilver, L., Perregrino, J.A. & Smith, C.J., Jr (1959) Estimation of hydrazine and three of its methyl derivatives. Amer. industr. Hyg. Ass. J., 20, 26

Harshman, R.C. (1957) The physical and chemical properties of alkyl hydrazines. Jet Propulsion, 27, 398

Juhász, J. (1967) On potential carcinogenicity of some hydrazine derivatives used as drugs. In: Truhaut, R., ed., Potential Carcinogenic Hazards from Drugs (UICC Monograph Series, Vol. 7), p. 180

Juhász, J., Baló, J. & Szende, B. (1966) Tumour-inducing effect of hydrazine in mice. Nature (Lond.), 210, 1377

Kelly, M.G., O'Gara, R.W., Yancey, S.T., Gadekar, K., Botkin, C. & Oliverio, V.T. (1969) Comparative carcinogenicity of N-isopropyl-α-(2-methyl-hydrazino)-p-toluamide · HCl (procarbazine hydrochloride), its degradation products, other hydrazines, and isonicotinic acid hydrazide. J. nat. Cancer Inst., 42, 337

McKennis, H., Jr, Yard, A.S., Weatherby, J.H. & Hagy, J.A. (1959) Acetylation of hydrazine and the formation of 1,2-diacetylhydrazine in vivo. J. Pharmacol. exp. Ther., 126, 109

Milia, U. (1965) Tumori polmonari da idrazina solfato somministrata a topi neonati del BALB/c/Cb/Se substrain. Lav. Ist. Anat. Univ. Perugia, 25, 73

Milia, U., Biancifiori, C. & Santilli, F.E.G. (1965) Late findings in pulmonary carcinogenesis by hydrazine sulphate in newborn BALB/c/Cb/Se substrain mice. Lav. Ist. Anat. Univ. Perugia, 25, 165

Mirvish, S.S., Chen, L., Haran-Guera, N. & Berenblum, I. (1969) Comparative study of lung carcinogenesis, promoting action in leukaemogenesis and initiating action in skin tumorigenesis by urethane, hydrazine and related compounds. Int. J. Cancer, 4, 318

Nair, V.R. & Nair, C.G.R. (1971) The titration of isoniazid and other hydrazine derivatives with chloramine-T. Analyt. Chim. Acta, 57, 429

Raphaelian, L.A. (1966) Hydrazine. In: Kirk, R.E. & Othmer, D.F., eds., Encyclopedia of Chemical Technology, 2nd ed., New York, John Wiley & Sons, Vol. 11, p. 164

Reed, R.A. (1957) Hydrazine and its derivatives. Roy. Inst. Chem. Lectures, Monographs and Reports, 5, 1

Roe, F.J.C., Grant, G.A. & Millican, D.M. (1967) Carcinogenicity of hydrazine and 1,1-dimethylhydrazine for mouse lung. Nature (Lond.), 216, 375

Severi, L. & Biancifiori, C. (1967) Cancerogenesi epatica nei topi CBA/Cb/Se e nei ratti Cb/Se da idrazina solfato. Epatologica, 13, 199

Severi, L. & Biancifiori, C. (1968) Hepatic carcinogenesis in CBA/Cb/Se mice and Cb/Se rats by isonicotinic acid hydrazide and hydrazine sulfate. J. nat. Cancer Inst., 41, 331

Smith, E.B. & Clark, D.A. (1972) Absorption of hydrazine through canine skin. Toxicol. appl. Pharmacol., 21, 186

Spinkova, V. (1971) Zur Bestimmung geringer Mengen Hydrazin in Isoniazid-Lösungen. Pharm. Acta helv., 46, 643

Toth, B. (1969) Lung tumor induction and inhibition of breast adenocarcinomas by hydrazine sulfate in mice. J. nat. Cancer Inst., 42, 469

Toth, B. (1971) Investigations on the relationship between chemical structure and carcinogenic activity of substituted hydrazines. Proc. amer. Ass. Cancer Res., 12, 55

Toth, B. (1972) Tumourigenesis studies with 1,2-dimethylhydrazine dichloride, hydrazine sulphate and isonicotinic acid in golden hamsters. Cancer Res., 32, 804

Wood, P.R. (1953) Determination of maleic hydrazide residues in plant and animal tissues. Analyt. Chem., 25, 1879

Yamamoto, R.S. & Weisburger, J.H. (1970) Failure of arginine glutamate to inhibit lung tumour formation by isoniazid and hydrazine in mice. Life Sci., 9, 285

1,1-DIMETHYLHYDRAZINE[*]

1. Chemical and Physical Data

1.1 Synonyms and trade names

Chem. Abstr. No.: 57-14-7

Asymmetrical-dimethylhydrazine; dimazine; dimethylhydrazine; N,N-dimethylhydrazine; UDMH; UNS-dimethylhydrazine; unsymmetrical-dimethylhydrazine

1.2 Chemical formula and molecular weight

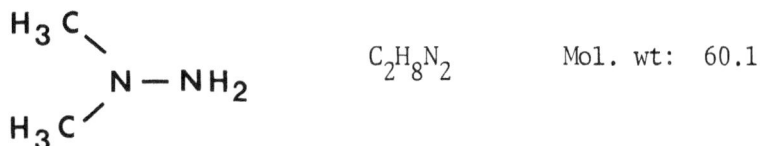

$$H_3C \diagdown \atop H_3C \diagup N - NH_2 \qquad C_2H_8N_2 \qquad \text{Mol. wt: } 60.1$$

1.3 Chemical and physical properties of the pure substance

(a) Description: A clear, colourless, flammable, hygroscopic liquid with the characteristic ammonia-like, fishy odour of alkyl hydrazines, fumes in air and gradually turns yellow

(b) Boiling-point: $63^{\circ}C$

(c) Melting-point: $-57^{\circ}C$

(d) Density: d_4^{22} 0.791

(e) Refractive index: $n_D^{22.3}$ 1.4075

(f) Solubility: Miscible with water, ethanol, ether, dimethyl-formamide and hydrocarbons

---

[*] Considered by the Working Group in Lyon, June 1973.

137

(g) Stability: The vapour is inflammable in air and ignites spontaneously when in contact with oxidizing agents. Solutions stored in the dark and the cold are relatively stable in the absence of oxidants (under certain conditions the compound may be highly explosive).

(h) Chemical reactivity: The compound is highly reactive. It is easily oxidizable and forms salts (e.g., the hydrochloride and oxalate).

(i) Volatility: The vapour pressure is 10 mm Hg at $-22^{o}C$ and 100 mm Hg at $16^{o}C$.

1.4 Technical products and impurities

1,1-Dimethylhydrazine (UDMH) is available in the United States as a single grade containing 98% (minimum) active ingredient and normally having 1.9% (maximum) dimethylamine and 0.3% (maximum) water content.

2. Production, Use, Occurrence and Analysis

A review on the properties, applications, reactions, storage and handling of UDMH has been published (FMC Corporation, 1972).

(a) Production and use[1]

Production of UDMH was first reported to the US Tariff Commission in 1956 (US Tariff Commission, 1957). Although this chemical can be made by the reaction of dimethylamine with chloramine (produced from ammonia and sodium hypochlorite), commercial production is believed to be based on the reduction of N-nitrosodimethylamine (which is made by the nitrosation of dimethylamine).

The US production rate for UDMH in early 1961 was estimated by one source to be in the range of 2.5-7.5 million kg per year (Liquid Propellant

---

[1] Data from Chemical Information Services, Stanford Research Institute, USA.

Information Agency, March 1961). Because only one company reports production of commercial quantities of UDMH, no data are published by the US Tariff Commission, but it is believed that current US production is less than 500 thousand kg per year.

UDMH is known to be made by one company in Japan, but no indication was found that it is produced in any other country.

By far the major use of UDMH to date has been as a storable, high-energy propellant for liquid-fuelled rockets. In the US, it was used in fuel for the Titan group of boosters used in the Gemini flights and in some of the Apollo flights. The Apollo fuel was reported to have been a 50:50 mixture of UDMH and hydrazine with nitrogen tetroxide as the oxidizer (Anon., 1969a).

Significant quantities of UDMH are used in the manufacture of N-dimethylaminosuccinamic acid, a plant growth regulator used in the US since 1963 to retard the growth of ornamentals such as chrysanthemums (Anon., 1965). This chemical probably has also found use in the last few years in the control of the vegetation, flowers or fruits of such crops as apples, grapes, peanuts, cherries, peaches and tomatoes (Anon., 1969b).

Another use of UDMH as a chemical intermediate, which may consume significant quantities in the future, was introduced in March 1973 by one US company manufacturing aminimides made from UDMH, and which offered for sale pilot plant quantities of eight of these products (Anon., 1973).

(b) Occurrence

UDMH has not been reported to occur as such in nature. It may be present in the waste streams from plants where it is produced or used. One source has reported that the burning of rocket fuels based on dimethyl-hydrazine and hydrazine produces exhaust gases which contain only trace quantities of unchanged fuel (Anon., 1969a). There is some evidence that N-nitrosodimethylamine might be formed during the burning of UDMH as a rocket fuel (Simoneit & Burlingame, 1971).

139

(c) Analysis

When 1,1-dialkylhydrazines are oxidized with mercury (II) sulphate
in aqueous sulphuric acid, one alkyl group splits off as formaldehyde, which
can be determined quantitatively. The lower detection limist is 0.2 μM
(Preussmann et al., 1968). Pinkerton et al.(1963) described a colorimetric
method for measuring microgram quantities of UDMH in blood, water and air
using trisodium pentacyanoaminoferroate as the colour reagent; readings are
taken spectrophotometrically at 500 nm. Milligram quantities can be estimated by potassium iodate titration (Jamieson, 1912; Feinsilver et al.,
1959). Bighi & Saglietto (1966) described a gas chromatographic method
for separating 1,1- and 1,2-dimethylhydrazines.

3. Biological Data Relevant to the Evaluation of Carcinogenic Risk to Man

3.1 Carcinogenicity and related studies in animals

(a) Oral administration

Mouse: Roe et al.(1967) gave by gavage daily doses of 0.5 mg
UDMH in water 5 days per week for 40 weeks to a group of 25 female
Swiss mice. Lung tumours were found in 1/8 mice (0.25 tumours/mouse*)
dying between 40 and 45 weeks and in 4/9 mice (2.6 tumours/mouse)
dying between 50 and 60 weeks. In 85 controls, 2/37 mice (0.05
tumours/mouse) and 6/42 mice (0.2 tumours/mouse) developed lung
tumours within the same periods.

Administration of 0.01% UDMH in the drinking-water of 50 male
and 50 female Swiss mice resulted in a high incidence of angiosarcomas
(79%), located in various organs. Besides these angiosarcomas, tumours
of the lungs (71%), kidneys (10%) and liver (6%) were observed. The
average latent period ranged from 42 to 61 weeks for the various
tumours (Toth, 1972, 1973).

---

* Multiplicity of tumours given as number of tumours/effective
number of mice.

Rat: Druckrey et al.(1967) reported that liver carcinomas developed in 3/20 BD rats given 70 mg UDMH in the drinking-water daily for life after 540, 806 and 1100 days (total doses, 36, 60 and 82 g/kg bw).

3.2 Other relevant biological data

(a) Animals

UDMH is absorbed rapidly through the skin of dogs and appears in the blood within 30 seconds (Smith & Clark, 1971).

3.3 Observations in man

Petersen et al.(1970) describe 6 cases of fatty liver associated with a rise in SGPT levels in 26 personnel working with liquid rocket fuels for up to 5 years.

4. Comments on Data Reported and Evaluation[1]

4.1 Animal data

1,1-Dimethylhydrazine (UDMH) is carcinogenic in mice after oral administration. The observation of a few liver tumours after high oral doses of UDMH occurring in rats after a long latent period does not allow a proper evaluation of the carcinogenic effect in this species.

4.2 Human data

No epidemiological data are available to the Working Group.

---

[1] See also the section "Extrapolation from animals to man" in the introduction to this volume.

5. References

Anon. (1965) Farm Chemicals, February, 32

Anon. (1969a) Chemical and Engineering News, July 14, 16

Anon. (1969b) Farm Chemicals, July, 19

Anon. (1973) Chemical Week, March 28, 26

Bighi, C. & Saglietto, G. (1966) The effect of alkalized adsorbent on the gas chromatographic separation of hydrazine derivatives. J. gas Chromat., 4, 303

Druckrey, H., Preussmann, R., Ivankovic, S. & Schmähl, D. (1967) Organotrope carcinogene Wirkungen bei 65 verschiedenen N-Nitroso-Verbindungen an BD Ratten. Z. Krebsforsch., 69, 103

FMC Corporation (1972) Dimazine[R] Product Bulletin, New York, Organic Chemicals Division

Feinsilver, L., Perregrino, J.A. & Smith, C.J., Jr (1959) Estimation of hydrazine and three of its methyl derivatives. Amer. industr. Hyg. Ass. J., 20, 26

Jamieson, G.S. (1912) A volumetric method for the determination of hydrazine. Amer. J. Science, 33, 352

Liquid Propellant Information Agency (March 1961) Liquid Propellant Manual, Unit 5, p. 10 (Available from Chemical Propulsion Information Agency, 8621 Georgia Avenue, Silver Springs, Maryland 20910, USA)

Petersen, P., Bredahl, E., Lauritsen, O. & Laursen, T. (1970) Examination of the liver in personnel working with liquid rocket propellant. Brit. J. industr. Med., 27, 141

Pinkerton, M.K., Lauer, J.M., Diamond, P. & Tamas, A.A. (1963) A colorimetric determination for 1,1-dimethylhydrazine (UDMH) in air, blood, and water. Amer. industr. Hyg. Ass. J., 24, 239

Preussmann, R., Hengy, H. & von Hodenberg, A. (1968) Eine neue photometrische Bestimmungsmethode von 1,1-Dialkylhydrazinen. Analyt. Chim. Acta, 42, 95

Roe, F.J.C., Grant, G.A. & Millican, D.M. (1967) Carcinogenicity of hydrazine and 1,1-dimethylhydrazine for mouse lung. Nature (Lond.), 216, 375

Simoneit, B.R. & Burlingame, A.L. (1971) Organic analyses of selected areas of Surveyor III recovered on the Apollo 12 mission. Nature (Lond.), 234, 210

Smith, E.B. & Clark, D.A. (1971) Absorption of unsymmetrical dimethyl-hydrazine (UDMH) through canine skin. Toxicol. appl. Pharmacol., 18, 649

Toth, B. (1972) Comparative studies with hydrazine derivatives. Carcino-genicity of 1,1-dimethylhydrazine, unsymmetrical (1,1-DMH) in the blood vessels, lung, kidneys, and liver of Swiss mice. Proc. amer. Assoc. Cancer Res., 13, 34

Toth, B. (1973) 1,1-Dimethylhydrazine (unsymmetrical) carcinogenesis in mice. Light microscopic and ultrastructural studies on neoplastic blood vessels. J. nat. Cancer Inst., 50, 181

US Tariff Commission (1957) Synthetic Organic Chemicals, United States Production and Sales, 1956, Second Series, Report No. 200

1,2-DIMETHYLHYDRAZINE[*]

1. Chemical and Physical Data

1.1 Synonyms and trade names

Chem. Abstr. No.: 54-07-3

N,N'-Dimethylhydrazine; symmetrical-dimethylhydrazine; hydrazomethane; SDMH

1.2 Chemical formula and molecular weight

$CH_3 - NH - NH - CH_3$ $C_2H_8N_2$ Mol. wt: 60.1

1.3 Chemical and physical properties of the pure substance

(a) Description: A clear, colourless, flammable, hygroscopic liquid with the characteristic ammonia-like, fishy smell of alkyl hydrazines

(b) Boiling-point: 80-81°C

(c) Melting-point: -9°C

(d) Density: d_4^{20} 0.8274

(e) Refractive index: n_D^{20} 1.4204

(f) Solubility: Miscible with water, ethanol, ether, dimethyl-formamide and hydrocarbons

(g) Stability: Solutions stored in the dark and the cold are relatively stable

(h) Volatility: The vapour pressure at -8°C is 10 mm Hg, and at 28°C it is 100 mm Hg

(i) Chemical reactivity: In the presence of traces of heavy metal ions, which act as catalysts, the compound is rapidly dehydrogenated to azomethane.

---

[*] Considered by the Working Group in Lyon, June 1973.

1.4 Technical products and impurities

No information is available to the Working Group, since 1,2-dimethyl-hydrazine (SDMH) is not produced commercially other than as a research chemical.

2. Production, Use, Occurrence and Analysis

(a) Production and use[1]

Although some United States producers of laboratory chemicals offer to make small quantities of this chemical on request, none produces it in commercial quantities. No information is available on the method of synthesis used by the laboratory chemical producers, but symmetrical dialkyl-hydrazines such as SDMH can be made by (i) reduction of the corresponding azine using lithium aluminium hydride, (ii) hydrolysis of alkyl-substituted diazacyclopropanes, or (iii) reaction of hydrazine with alkyl halides (Raphaelian, 1966).

No indication was found that SDMH is produced in countries other than the US.

There are no known commercial uses for SDMH, although it was evaluated experimentally as a high-energy rocket fuel.

(b) Occurrence

SDMH has not been reported to occur as such in nature.

(c) Analysis

Oxidation of SDMH with mercury (II) sulphate in aqueous sulphuric acid splits off one alkyl group as formaldehyde, which is released and can be determined quantitatively. The lower detection limit is 0.1-0.2 µM (Preussmann et al., 1968). Milligram quantities can be determined by potassium iodate titration or potentiometric and colorimetric methods

---

[1] Data from Chemical Information Services, Stanford Research Institute, USA.

(Jamieson, 1912; Feinsilver et al., 1959). SDMH can be separated from 1,1-dimethylhydrazine by gas chromatography (Bighi & Saglietto, 1966).

3. Biological Data Relevant to the Evaluation of Carcinogenic Risk to Man

3.1 Carcinogenicity and related studies in animals

(a) Oral administration

<u>Mouse</u>: Toth & Wilson (1971) gave SDMH dihydrochloride in the drinking-water to Swiss mice for life. The daily intake was approximately 58 µg for females and 87 µg for males. After median latent periods of 45 and 42 weeks, 49 females (98%) and 46 males (92%) showed angiosarcomas, mainly localized in the muscle, liver and pararenal tissues. In addition, 22 females (44%) and 12 males (24%) had lung adenomas after latent periods of 49 and 44 weeks, respectively. In 110 female and 110 male untreated controls, 4% of the females and 2% of the males had angiosarcomas, and 13% of the females and 10% of the males had lung tumours.

<u>Rat</u>: Druckrey et al.(1967) and Druckrey (1970) reported the selective action of a neutralized solution of SDMH-dihydrochloride on the intestine of the rat. Oral doses readily produced cancers of the intestine, i.e., adenocarcinomas, mostly of the colon and rectum. Doses of 21 mg/kg bw/week for 11 weeks by stomach tube produced intestinal tumours in 13/14 BD strain rats, with a mean induction time of about 200 days. A lower dose of SDMH (3 mg/kg bw) in the drinking-water for 5 days per week produced a high incidence of malignant haemangioendotheliomas of the liver with multiple metastases in the lungs, but no intestinal tumours (Druckrey, 1970).

Pozharissky (1972) treated a group of 63 male outbred rats weekly with single oral doses of 21 mg/kg bw (total dose, 573 mg/kg bw). In all, 26/27 animals developed tumours within 119 to 234 days. Tumours of the large and small intestine, appendix and anal region were observed. About 50% of the tumours produced metastases.

Hamster: Groups of 50 male and 50 female hamsters received average daily intakes of 0.16 mg SDMH in the drinking-water for life. Angiosarcomas of the blood vessels occurred in 82% of males and 89% of females, with an average latent period of 52 weeks. The frequencies of these lesions in the various tissues were in the following order: liver, lung, muscle, heart and pancreas. In addition, tumours of the caecum, mainly polypoid adenomas, but some adenocarcinomas (and also in the females some leiomyosarcomas), developed in 12% of the males and 34% of the females in about 60 weeks. Tumours of the liver were found in 14% of the males and 20% of the females (Toth, 1972).

(b)   Subcutaneous and/or intramuscular administration

Mouse: A group of 90 mice, given s.c. doses of 15 mg/kg bw/ week of SDMH-dihydrochloride for 22 weeks, developed tumours in the descending colon and rectum in 52/58 survivors (Hawks et al., 1971/ 1972). In further studies, 19/22 males given 10 mg/kg bw/week and 22/28 females given 15 mg/kg bw/week developed squamous cell carcinomas in the anal region and adenocarcinomas in the descending colon and rectum within 24 weeks (Pegg & Hawks, 1971). In a group of 110 mice, polyps and/or carcinomas developed in the large intestine of most animals between 180 and 240 days after total doses of 420-540 mg/kg bw given by weekly s.c. injections of 15 mg/kg bw (Wiebecke et al., 1969). Weekly s.c. injections of 20 mg/kg bw SDMH-dihydrochloride for 2 to 24 weeks to 34 CF_1 mice produced multiple carcinomas of the colon in 90% of the animals. The first tumour was detected in a mouse dying at 135 days (Thurnherr et al., 1973).

Rat: A number of studies have demonstrated the carcinogenicity of SDMH by the subcutaneous route in several rat strains (Druckrey et al., 1967; Druckrey, 1970; Martin et al., 1973; Preussmann et al., 1969). Weekly injections of 21 or 7 mg/kg bw for 36 weeks or 263 days, respectively, produced a high incidence of adenocarcinomas of the colon, rectum and duodenum within 184 or 333 days, respectively. Tumours of the small intestine and liver were found only at the higher dose level (Druckrey, 1970). Pozharissky (1972) produced a 100% incidence of intestinal tumours after total doses of 328-822 mg/kg bw.

The average time at which tumours were found was dose related, being shorter at higher doses (range, 116 to 393 days).

Hamster: Male golden hamsters were injected intramuscularly once weekly with doses of 4.3 mg/kg bw. In 14/25 survivors, receiving a total dose of 146.2 mg/kg bw, 5 hepatocellular carcinomas, 1 carcinoma of the stomach and 8 adenocarcinomas of the small and large intestine were observed. The median induction time was 261 ± 49 days (Osswald & Krüger, 1969).

3.2 Other relevant biological data

(a) Animals

SDMH is probably not carcinogenic per se but is activated in vivo by metabolic processes to form the ultimate carcinogen. Oxidative dealkylation of SDMH would yield reactive intermediates which could act as methylating agents in vivo, as demonstrated by the methylation of RNA in the colon after administration of ^{14}C SDMH to mice (Hawks et al., 1972).

(b) Carcinogenicity of metabolites

A possible metabolite of SDMH, azoxymethane, produced a high yield of carcinomas of the colon and the rectum in rats when given orally or when injected subcutaneously, and liver tumours when given at lower doses (Preussmann et al., 1969). Similar results were obtained by Ward et al. (1973) in mice given i.p. injections, or in rats given s.c. injections, of azoxymethane. Tumours appeared within 5 to 6 months.

Gennaro et al. (1973) obtained tumours in segments of colon surgically transposed to the small intestine in 18 rats injected i.m. weekly with 8 mg/kg bw azoxymethane. Segments of small intestine transposed to the colon in 17 rats given azoxymethane failed to develop tumours, although tumours did appear in the colon.

A single i.v. injection of 20 mg/kg bw azoxymethane to BD rats at the 22nd day but not at the 15th day of gestation resulted in the occurrence of renal and neurogenic tumours in the offspring (Druckrey, 1973).

3.3 Observations in man

No data are available to the Working Group.

4. Comments on Data Reported and Evaluation[1]

4.1 Animal data

1,2-Dimethylhydrazine (SDMH) is carcinogenic in mice, rats and hamsters following oral and subcutaneous or intramuscular administration.

4.2 Human data

No epidemiological data are available to the Working Group.

---

[1] See also the section "Extrapolation from animals to man" in the introduction to this volume.

5. References

Bighi, C. & Saglietto, G. (1966) The effect of alkalized adsorbent on the gas chromatographic separation of hydrazine derivatives. J. gas Chromat., 4, 303

Druckrey, H. (1970) Production of colonic carcinomas by 1,2-dialkylhydrazines and azoxyalkanes. In: Burdette, W.J. ed., Carcinomas of the Colon and Antecedent Epithelium, Springfield, Illinois, Thomas, p. 267

Druckrey, H. (1973) Chemical structure and action in transplacental carcinogenesis and teratogenesis. In: Tomatis, L. & Mohr, U., eds., Transplacental Carcinogenesis, Lyon, IARC Scientific Publications, 4, p. 45

Druckrey, H., Preussmann, R., Matzkies, F. & Ivankovic, S. (1967) Selektive Erzeugung von Darmkrebs bei Ratten durch 1,2-Dimethylhydrazin. Naturwissenschaften, 54, 285

Feinsilver, L., Perregrino, J.A. & Smith, C.J., Jr (1959) Estimation of hydrazine and three of its methyl derivatives. Amer. industr. Hyg. Ass. J., 20, 26

Gennaro, A.R., Villanueva, R., Sukonthaman, Y., Vathanophas, V. & Rosemond, G.P. (1973) Chemical carcinogenesis in transposed intestinal segments. Cancer Res., 33, 536

Hawks, A., Farber, E. & Magee, P.N. (1971/1972) Equilibrium centrifugation studies of colon DNA from mice treated with the carcinogen 1,2-dimethylhydrazine. Chem.-biol. Interactions, 4, 144

Hawks, A., Swann, P.F. & Magee, P.N. (1972) Probable methylation of nucleic acids of mouse colon by 1,2-dimethylhydrazine in vivo. Biochem. Pharmacol., 21, 432

Jamieson, G.S. (1912) A volumetric method for the determination of hydrazine. Amer. J. Science, 33, 352

Martin, F., Michiels, R., Bastien, H., Justrabo, E., Bordes, M. & Viry, B. (1973) An experimental model for cancer of the colon and rectum. Digestion, 8, 22

Osswald, H. & Krüger, F.W. (1969) Die cancerogene Wirkung von 1,2-Dimethylhydrazin beim Goldhamster. Arzneimittel-Forsch., 19, 1891

Pegg, A.E. & Hawks, A. (1971) Increased transfer ribonucleic acid methylase activity in tumours induced in the mouse colon by administration of 1,2-dimethylhydrazine. Biochem. J., 122, 121

Pozharissky, K.M. (1972) Intestinal tumours induced in rats with 1,2-dimethylhydrazine. Vop. Onkol., 18, 64

Preussmann, R., Druckrey, H., Ivankovic, S. & von Hodenberg, A. (1969) Chemical structure and carcinogenicity of aliphatic hydrazo, azo and azoxy compounds and of triazenes, potential in vivo alkylating agents. Ann. N.Y. Acad. Sci., 163, 697

Preussmann, R., Hengy, H., Lübbe, D. & Von Hodenberg, A. (1968) Photometrische Bestimmung aliphatischer Azo- und Hydrazo-Verbindungen. Analyt. Chim. Acta, 41, 497

Raphaelian, L.A. (1966) Hydrazine. In: Kirk, R.E. & Othmer, D.F., eds., Encyclopedia of Chemical Technology, 2nd ed., New York, John Wiley & Sons, Vol. 11, p. 164

Thurnherr, N., Deschner, E., Stonehill, E.H. & Lipkin, M. (1973) Induction of adenocarcinomas of the colon in mice by weekly injections of 1,2-dimethylhydrazine. Cancer Res., 33, 940

Toth, B. (1972) Tumorigenesis studies with 1,2-dimethylhydrazine dihydrochloride, hydrazine sulfate and isonicotinic acid in golden hamsters. Cancer Res., 32, 804

Toth, B. & Wilson, R.B. (1971) Blood vessel tumorigenesis by 1,2-dimethylhydrazine dihydrochloride (symmetrical). Amer. J. Path., 64, 585

Ward, J.M., Yamamoto, R.S., Benjamin, T., Brown, C.A. & Weisburger, J.H. (1973) Experimental colon cancer in rats and mice. J. Amer. vet. med. Ass. (in press)

Wiebecke, B., Löhrs, U., Gimmy, J. & Eder, M. (1969) Erzeugung von Darmtumoren bei Mäusen durch 1,2-Dimethylhydrazin. Z. ges. exp. Med., 149, 277

1,2-DIETHYLHYDRAZINE[*]

1. Chemical and Physical Data

1.1 Synonyms and trade names

Chem. Abstr. No.: 1615-80-1

N,N'-Diethylhydrazine; symmetrical-diethylhydrazine; hydrazoethane; SDEH

1.2 Chemical formula and molecular weight

$CH_3CH_2NHNHCH_2CH_3$ $C_4H_{12}N_2$ Mol. wt: 88.2

1.3 Chemical and physical properties

(a) Description: Colourless liquid

(b) Boiling-point: 85-86°C

(c) Density: d_4^{20} 0.797

(d) Refractive index: n_D^{20} 1.4204

(e) Solubility: The hydrochloride is soluble in water and ethanol and slightly soluble in acetone

(f) Stability: The free base is highly reactive and oxidizes readily

(g) Chemical reactivity: In the presence of trace amounts of heavy metal ions, which act as catalysts, the compound is rapidly dehydrogenated to azoethane.

1.4 Technical products and impurities

No information is available to the Working Group since 1,2-diethyl-hydrazine (SDEH) is not produced commercially other than as a research chemical.

---

[*] Considered by the Working Group in Lyon, June 1973.

2. Production, Use, Occurrence and Analysis

(a) Production and use[1]

There are no known commercial uses for SDEH, although it has been evaluated experimentally as a high-energy rocket fuel. The compound is used in chemical laboratories, e.g., for the synthesis of symmetrical di-Mannich-bases (Ried & Wesselborg, 1958). SDEH may be produced during the thermic decomposition of ethylamine (Taylor & Ditman, 1936).

(b) Occurrence

SDEH has not been reported to occur as such in nature.

(c) Analysis

Oxidation of 1,2-dialkylhydrazines with mercury (II) sulphate in aqueous sulphuric acid splits off one alkyl group as acetaldehyde, which is released and can be determined quantitatively. The lower detection limit is 0.1-0.2 µM (Preussmann et al., 1968).

3. Biological Data Relevant to the Evaluation of Carcinogenic Risk to Man

3.1 Carcinogenicity and related studies in animals

(a) Subcutaneous and/or intramuscular administration

Rat: Rats of several BD-strains were injected s.c. once weekly with 25, 50 or 100 mg/kg bw (calculated as base) of a neutralized aqueous solution of SDEH-dihydrochloride for 30 weeks. The total doses were 770, 1400 and 2700 mg/kg bw. Tumours in the brain, olfactory bulbs, mammary glands and liver were seen in 43/45 rats. The average latent periods ranged from 250 days at the lowest dose to 215 days at the highest dose (Druckrey et al., 1966).

---

[1] Data from Chemical Information Services, Stanford Research Institute, USA.

(b) Other experimental systems

Pre-natal exposure: Druckrey et al.(1968) gave single i.v. injections of 50 and 150 mg/kg bw SDEH to pregnant BD rats on the 15th day of gestation. Of the offspring 18/19 and 11/12, respectively, died between the 126th and 498th day with tumours of the brain, the spinal cord or the peripheral nervous system. This finding was subsequently confirmed in a large number of animals (Druckrey, 1973).

3.2 Other relevant biological data

(a) Animals

1,2-Diethylhydrazine (SDEH) is probably not carcinogenic per se but is activated in vivo by metabolic processes to form the ultimate carcinogen.

(b) Carcinogenicity of metabolites

The possible metabolites of SDEH, azoethane and azoxyethane have been shown to be carcinogenic (Druckrey, 1968a; Druckrey et al., 1965; Druckrey et al., 1967; Druckrey, 1968b; Preussmann et al., 1969). Azoethane, in s.c. injections of 50 mg and 100 mg/kg bw once weekly for lifetime or for 13 weeks to a total of 22 BD rats, induced tumours in all animals. The tumours were localized in the haematopoietic system, liver, forestomach, nasal cavity and brain (Druckrey et al., 1965). A single i.v. injection of 50 mg/kg bw azoxyethane on the 15th day of gestation and inhalation exposure of an evaluated dose of 37-150 mg/kg bw azoethane on the 15th or 22nd day to BD rats resulted in a high incidence of neurogenic tumours in the offspring (Druckrey, 1973).

3.3 Observations in man

No data are available to the Working Group.

4. Comments on Data Reported and Evaluation[1]

4.1 Animal data

1,2-Diethylhydrazine (SDEH) is carcinogenic in rats by subcutaneous administration and transplacental exposure, the only species and routes tested.

4.2 Human data

No epidemiological data are available to the Working Group.

-----

[1] See also the section "Extrapolation from animals to man" in the introduction to this volume.

5. References

Druckrey, H. (1968a) Chemical constitution and carcinogenic activity of alkylating substances. In: Weber, K.H., ed., Alkylierend wirkende Verbindungen. Zweite Konferenz über aktuelle Probleme der Tabakforschung, Freiburg, 1967, Hamburg, Wissenschaftliche Forschungsstelle im Verband der Cigarettenindustrie, p. 42

Druckrey, H. (1968b) Organotropic carcinogenesis and mechanisms of action of symmetrical dialkylhydrazines and azo- and azoxyalkanes. Fd Cosmet. Toxicol., 6, 578

Druckrey, H. (1973) Chemical structure and action in transplacental carcinogenesis and teratogenesis. In: Tomatis, L. & Mohr, U. eds., Transplacental Carcinogenesis, Lyon, IARC Scientific Publications, 4, p. 45

Druckrey, H., Ivankovic, S., Preussmann, R., Landschütz, C., Stekar, J., Brunner, U. & Schagen, B. (1968) Transplacentar induction of neurogenic malignomas by 1,2-diethyl-hydrazine, azo-, and azoxyethane in rats. Experienta (Basel), 24, 561

Druckrey, H., Preussmann, R., Ivankovic, S., Schmidt, C.H., So, B.T. & Thomas, C. (1965) Carcinogene Wirkung von Azoäthan und Azoxyäthan an Ratten. Z. Krebsforsch., 67, 31

Druckrey, H., Preussmann, R., Ivankovic, S., Schmähl, D., Afkham, J., Blum, G., Mennel, H.D., Müller, M., Petropoulus, P. & Schneider, H. (1967) Organotrope carcinogene Wirkungen bei 65 verschiedenen N-Nitroso-Verbindungen an BD-Ratten. Z. Krebsforsch., 69, 103

Druckrey, H., Preussmann, R., Matzkies, F. & Ivankovic, S. (1966) Carcinogene Wirkung von 1,2-Diäthylhydrazin an Ratten. Naturwissenschaften, 53, 557

Preussmann, R., Druckrey, H., Ivankovic, S. & von Hodenberg, A. (1969) Chemical structure and carcinogenicity of aliphatic hydrazo, azo, and azoxy compounds and of triazenes, potential in vivo alkylating agents. Ann. N.Y. Acad. Sci., 163, 697

Preussmann, R., Hengy, H., Lübbe, D. & von Hodenberg, A. (1968) Photometrische Bestimmung aliphatischer Azo- und Hydrazo-Verbindungen. Analyt. Chim. Acta, 41, 497

Ried, W. & Wesselborg, K. (1958) Beiträge zur Mannich-Reaktion. II. Zweistufen-Verfahren zur Darstellung symmetrischer Di-Mannich-Basen. Justus Liebigs Ann. Chem., 611, 71

Taylor, H.A. & Ditman, J.G. (1936) The decomposition of ethylamine and diethylhydrazine. J. chem. Phys., 4, 212

ISONICOTINIC ACID HYDRAZIDE*

1. Chemical and Physical Data

1.1 Synonyms and trade names

Chem. Abstr. No.: 54-85-3

Isonicotinoylhydrazine; isonicotinylhydrazine; 4-pyridinecarboxylic acid hydrazide; pyridine-4-carboxylic acid hydrazide; pyridine-γ carboxylic acid hydrazide

5015 RP, Actobine, Actotibine, Aldoxal, Andrazida, Aneufos simple, Anteben, Antimicina, Antituberkulosum "Trogalen", Armazide, Azuren, Bacikoch, Bacillin, Banzid, Basidracida, Becazida, Buco-Hidracid, Cedin, Cemidón, Chemiazid, Chemidon, Cin-Vis, Cotinazin, Cotinizin, Dardex, Diazid, Dif-Azidrin, Dinacrin, Dioforin, Ditubin, Dracotil, Ducun, Ebidene, Eralon, Ertuban, Eutizon, Evalon, Evigril, Fimalene, Fimazid, FSR-3, GINK, Hain, HIA, Hidraciber, "Hidracida", Hidrafasa, Hidral-Grey, Hidralter, Hidranic, Hidranizil, Hidra-Noxi, Hidraquimia, Hidrasegur, Hidrasix, Hidrasonil, Hidrassal, Hidrastol, Hidratorax, Hidrazida-I.N., Hidrazinol, Hidrina, Hidrulta, Hidrun, Hiperazida, Hycozid, Hydra, Hydrazid "Polfa", Hydrazide "Horus", Hydrazide "Médial", Hydrazide "Otsuka", Hydrozin, Hyzyd, Idocobin, Ido-Tebin, Idrazil, INA, INAH, Inazid, INH, Inhizid, Inizid, Iscotin, Isidrina, Ismazide, Isobicina, Isocid, Isocidene, Isocotin, Iso-Dexter, Isolyn, Isomerina, Isonazida, Isonerit, Isonex, Isonhidrol, Isoniazid, Iso- niazide, Isoniazone, Isonicazide, Isonicid, Isonico, Isonicol, Iso- nicotan, Isonicotil, Isonide, Isonidrin, Isonikazid, Isonilex, Iso- nilyd, Isonin, Isonindon, Isoniton, Isonitrit, Isonizan, Isonizide, Iso-Pentabine, Isopyrin "Astra", Isotebe, Isotebezid, Isotinyl, Iso- tubin, Isoxine, Isozid, Isozide, Isozin, Isozyd, Isonisin, Kemia, Kozidrina, Kridansimple, L 1945, Laniazid, Laniozid, Lefos, Leotubrin- Wirkstoff, Lesviazida, Lubacida, Marsilid, Marvidrazida, Mesegacida,

---

* Considered by the Working Group in Lyon, June 1973.

Micosan, Milazide, Mybasan, Mycoseptina, Neoplasina, Neoteben, Neoxin, Neoxon, Neumandin, Nevin, Niadrin, Niazid, Niazida, Nicazid, Nicazide, Nicetal, Nicizina, Niconyl, Nicosciorin, Nicotibina, Nico- tibine, Nicotisan, Nicotubin, Nicozid, Nicozide, Nicozin, Nidaton, Nidrazid, Nieteban, Nikozid, Niplen, Nitadon, Nortibina, Nydrazid, Nyscozid, Pacrazid, Pelazid, Percin, Percitron, Peritracida, Phthisen, Pycazide, Pyreazid, Pyricidin, Pyridicin, Pyrinal, Pyrizidin, Raumanon, Rebilon, Retozide, Rimicid, Rimifon, Rimitsid, Rinverfons, Roberazyd, Robisellin, Roxifen "Miquel", RU-EF-Tb, Sanohidrazina, Sauterazid, Stanozide, Sumifon, Tb-Phlogin, T.B. Razide, TB-Vis, Tebecid, Tebecin, Tebenic, Tebesium-Wirkstoff, Tebetracin, Tebexin, Tebilon, Tebos, Teebaconin, Tekazin, Tibazide, Tibemid, Tibinid, Tibinide, Tibison, Tibitan, Tibivis, Tibizide, Tibusan, Tisin, Tisiodrazida, Tizide, Tizopan, Tubazid, Tubazide, Tubeco, Tubercid, Tuberian, Tuberon, Tubezin, Tubicon, Tubilysin, Tubomel, Tubriz, Tyvid, Unicozyde, Vaza- drine, Vederon, Zidafimia, Zideluy, Zinadon, Zonazide

1.2 Chemical formula and molecular weight

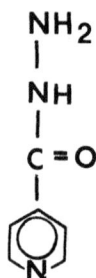

$$\begin{array}{c} NH_2 \\ | \\ NH \\ | \\ C = O \\ | \\ \end{array}$$

$C_6H_7N_3O$ Mol. wt: 137.2

1.3 Chemical and physical properties of the pure substance

(a) Description: Crystalline solid

(b) Melting-point: 171.4^{o}C

(c) Absorption spectroscopy: λ max 262-264 nm; log ε = 4140-4520

(in water)

(d) Solubility: Soluble in water (12% at 20°C, 25% at 37°C), methanol, ethanol and methyl ethyl ketone; sparingly soluble in acetone and chloroform; insoluble in carbon tetrachloride, ether and benzene

(e) <u>Stability</u>: Stable at room temperature for more than 14 days in aqueous solution and more than 6 weeks when stored at about $4^{\circ}C$

(f) <u>Chemical reactivity</u>: Can react as a weak acid or a weak base; can be hydrolyzed to release hydrazine and can be destroyed by oxidative and reductive reactions at the hydrazine moiety (Krüger-Thiemer, 1956; Scott, 1952; Canbäck, 1952).

1.4 Technical products and impurities

Isonicotinic acid hydrazide (INH) is available in a USP grade containing a minimum of 98% of the active ingredient. Small amounts of isomeric compounds and hydrazine may be present.

2. Production, Use, Occurrence and Analysis

(a) <u>Production and use</u>[1]

INH was first synthesized in the laboratory in 1912. It is produced commercially by the oxidation of 4-methylpyridine or 4-ethylpyridine to isonicotinic acid, which is reacted as the methyl or ethyl ester with hydrazine or hydrazine hydrate. Some INH may also be prepared by the reaction of hydrazine with 4-cyanopyridine (Merck & Co. Inc., 1968).

The capacity of the only United States producer of INH is not known. The total US production volume for a group of seven antileprotic and anti-tubercular agents (including INH) was reported to have been 100 thousand kg in 1970 (US Tariff Commission, 1972). INH and p-aminosalicyclic acid are believed to have constituted nearly all of this production volume.

The annual production in India (by nine manufacturers) ranged from 62 thousand kg in 1964 to 32 thousand kg in 1968. A new plant with a capacity of 118 thousand kg was started up in 1971.

---

[1] Data from Chemical Information Services, Stanford Research Institute, USA.

In Italy, three manufacturers produced about 400 thousand kg per annum between 1965 and 1968. In 1969, the annual capacity was recorded as 270 thousand kg.

The production of INH in Japan amounted to 180 thousand kg in 1965. The capacity of one plant in 1971 was reported to be 460 thousand kg per year.

In 1967, the Federal Republic of Germany was reported to have two producers of INH; and Denmark and Sweden were each reported to have one producer in 1970. Production and capacity data are not reported for these countries.

A small amount of INH may be used in the synthesis of isoniazid methane sulphonate and its sodium and calcium salts (all of which are anti-tubercular agents) and in the synthesis of nialamid (isonicotinic acid 2-(2-(benzylcarbamoyl)-ethyl)hydrazide), an antidepressant; but its major use is as a drug in its own right.

INH has been used almost exclusively as an antitubercular agent since 1952 (Peters et al., 1965). Although it reportedly has been used in veter-inary medicine in the past (e.g., in the treatment of tuberculosis in mink), no evidence was found that it is presently being used for this purpose. It is sold as a prescription drug in the form of tablets, as an injectable preparation and as a syrup. It is also sold, in the US and Japan at least, in mixtures with other drugs. A typical human dosage level is 300 mg daily (taken orally or intramuscularly) both in the treatment of active tubercu-losis and in preventive therapy. It may be used in combination with other antitubercular agents (streptomycin, ethambutol hydrochloride, and sodium p-aminosalicylate) to prevent the development of drug-resistant bacterial strains.

Based on the average sales price and the total sales value, US con-sumption of this substance in 1971 is estimated to have been about 10 thousand kg. This quantity is far less than the apparent US supply in 1971 (imports of 53 thousand kg in 1971 plus domestic production), but the reason for the discrepancy could not be determined from the available statistics.

Imports of INH by other countries have been reported as follows: Brazil - 45 thousand kg (1970); Republic of South Africa - 63 thousand kg (1969).

(b) Occurrence

INH has not been reported to occur as such in nature. It may occur in waste streams from plants where it is made or used as a chemical intermediate.

(c) Analysis

Many methods for the quantitative estimation of INH are known (Scott, 1952; Canbäck, 1952; Gemeinhard & Rangnick, 1953; Kelly & Poet, 1952; Ellard et al., 1972; Krüger-Thiemer, 1956; Vivien et al., 1972) including iodometric, bromometric and argentometric methods (Kern, 1958), a vanadametric method (Rao & Rao, 1971a), a potentiometric determination (Rao & Rao, 1971b), a chromatographic determination (Leuschner, 1954; Zamboni & Fachinelli, 1954), a gas chromatographic determination (Finkle et al., 1971), colour reactions (Vivien et al., 1972) and using chloramine-T (Nair & Nair, 1971). Methods for the determination of INH and its metabolites in human urine and in blood have been described (Prescott et al., 1953; Peters et al., 1965; Dymond & Russell, 1970).

3. Biological Data Relevant to the Evaluation of Carcinogenic Risk to Man

3.1 Carcinogenicity and related studies in animals

(a) Oral administration

Mouse: Several studies on different strains have demonstrated the carcinogenicity of INH by this route of administration. Mori & Yasuno (1959) and Mori et al.(1960) administered diets containing 0.25, 0.125, 0.1, 0.06 or 0.01% INH to 'dd' mice. The pulmonary tumour incidences were 100, 70, 60, 50 and 8%, respectively, in mice surviving 7 months from the beginning of treatment. Other groups of 20 mice which were fed 0.25% INH in the diet for periods of 2,3,4 and 7 months had corresponding tumour incidences of 40, 50, 50 and 100%. The controls had a negative incidence rate in one group, and in another

163

group the pulmonary tumour incidence was 3%.

Similar results were found in various strains of mice by Biancifiori & Ribacchi (1962a,b), Biancifiori et al.(1963), Ribacchi et al.(1963), Weinstein & Kinosita (1963), Toth & Shubik (1966a,b), Kelly et al.(1969), Toth & Toth (1970), Yamamoto & Weisburger (1970), Jones et al.(1971) and Linnik (1972). The subject was reviewed by Biancifiori & Severi (1966).

Rat: Studies on the carcinogenicity of INH given orally to rats have shown few positive effects.

Loscalzo (1964) gave 30 mg INH/kg bw in the drinking-water daily for 290 or 355 days to 60 albino rats. No tumours were found. Peacock & Peacock (1966), Lucchesi et al.(1967) and Toth & Toth (1970) could not demonstrate any tumourigenic action of INH in extensive, long-term tests in several strains of rats. Severi & Biancifiori (1968) gave Cb/Se rats daily doses of about 35 mg INH in the drinking-water for 48 weeks. Of the males, 1/49 developed a liver tumour and 2/49 had lung tumours. Of the females, 11/40 had fibroadenomas of the mammary gland. No liver, lung or mammary tumours were found in 28 male and 22 female controls of which 19 males and 14 females were killed at 104 weeks.

Hamster: Peacock & Peacock (1966) failed to produce pulmonary tumours in a group of 7 male and 9 female hamsters given 0.25% INH in the drinking-water for life (up to 90 weeks; total dose, 1.3-7.9 g). One hepatoma and 1 haemangioma were found in 2 females, but no control data are available. Concentrations of 0.1 or 0.2% INH administered in the drinking-water to groups of 50 male and 50 female Syrian golden hamsters for life, or 0.3% INH in the drinking-water to 35 male and 36 female hamsters for 42 weeks (with observation up to 104 weeks) failed to produce a significant carcinogenic effect (Toth & Boreisha, 1969; Toth & Shubik, 1969).

(b) Subcutaneous and/or intramuscular administration

Mouse: Mori et al.(1960) injected 'dd' mice s.c. with 2 mg

164

INH every 2 days for 18 weeks. The incidence of pulmonary tumours
(adenomas) in 15 animals surviving 7 months was 53% in the experimental group and 11% in the 9 control animals.

Jones et al.(1971) injected Strong A and BALB/c mice s.c. with
2 mg INH/animal on alternate days for 18 weeks. The lung tumour incidence was 52% (51/99) in males and 61% (58/95) in females of the
Strong A mice and 39% (28/71) in males and 40% (21/53) in females of
the BALB/c mice. A lower dose, 0.4 mg INH, administered in the same
way did not increase the lung tumour incidence in Strong A mice.
Saline-treated controls had lung tumours in 39% (81/210) Strong A
males and in 36% (76/209) females, and in 31% (32/102) BALB/c males
and in 22% (20/90) females.

(c) Intraperitoneal administration

Mouse: Juhász et al.(1957) administered 30 doses of INH (total
dose,82 mg per animal) to white mice within 3 months. In 7.5 months
tumours developed in 14/45 mice, including 7 lung adenomas, 6 leukaemias and 1 reticulum cell sarcoma of the liver. No tumours were observed in 50 control mice. With total doses of 40-55 mg per mouse,
tumours (mainly mediastinal lymphosarcomas and myeloid leukaemias)
developed in 15/50 mice within 13 months. Only 1/50 control mice
developed a myeloid leukaemia (Juhász et al., 1963).

Schwan (1962) noted lung tumours in 17/45 R_3-strain mice after
daily doses of 1.25 mg INH per mouse administered for 87 days, starting when the mice were 6 months of age. No lung tumours were seen
in a similar group of 45 control mice, nor in 380 controls killed at
various ages.

3.2 Other relevant biological data

(a) Animals

Colvin (1969) presented a review of the metabolism of hydrazines,
including INH. The in vivo metabolism of INH in the rabbit and the in
vitro metabolism in some organs and tissues of the rat and ox (such as
rat brain, kidney, muscle, liver and myocardic tissue and ox kidney)

165

yields isonicotinic acid and ammonia, the latter being derived from a rapid breakdown of the hydrazine group (Porcellati & Preziosi, 1954). N-Acetyltransferase from mammalian liver catalyzes the acetylation of INH (Steinberg et al., 1971). INH is converted to hydrazine and isonicotinic acid by a resistant strain of Mycobacterium avium (Toida, 1962).

(b) Man

Peters et al.(1965) found the most important metabolites of INH in urine to be 1-acetyl-2-isonicotinoylhydrazine (acetyl-INH), N-acetyl-N'-isonicotinic acid, isonicotinylglycine, pyruvic acid isonicotinylhydrazone and α-oxoglutaric acid isonicotinylhydrazone (see also Krüger-Thiemer, 1957; Iwainsky, 1965). Acetyl-INH is converted in serum to isonicotinic acid and acetylhydrazine (Ellard et al., 1972). Vivien et al.(1972) stated that INH is metabolized in man without the production of hydrazine. The extent of acetylation of INH can vary among individuals because of the genetic polymorphism of the corresponding N-acetyltransferase (Schulte et al., 1970; Gelber et al., 1971).

(c) Comparison of animal and human data

Hughes (1953) reported 1-acetyl-2-isonicotinoylhydrazine to be the major excretory product of INH in rhesus monkeys and in man, but no splitting of INH into isonicotinic acid and hydrazine could be demonstrated. Smith (1966) and Frohberg (1970) considered the metabolic behaviour of INH in monkeys and in man to be identical, but to be different in monkeys and dogs.

(d) Carcinogenicity of metabolites

A concentration of 0.4% 1-acetyl-2-isonicotinoylhydrazine given daily in the drinking-water of 50 male and 50 female Swiss mice for life resulted in the development of lung tumours (mainly adenomas, but some adenocarcinomas) in 58% of the male mice and 77% of the female mice. The incidence of lung tumours in controls was 10% and 12% in the males and females, respectively. The average latent period for tumour production was 65 and 79 weeks in the male and female treated animals (Toth & Shimizu, 1973).

3.3 Observations in man

Hammond et al.(1967), using data from a large prospective study of lung cancer in a population of more than 1 million, showed an excess of deaths due to lung cancer (45 cases versus 36.5 expected) among 18,963 tuberculous patients, 10% of whom had received INH treatment. The excess was not significant at the P 0.05 level. In another group of 311 patients treated with INH between 1951 and 1956, 10 lung cancer deaths had been reported by 1966, whereas 6.3 were expected. This difference was not significant at the P 0.05 level. A further group of 502 pregnant women, observed between 1953 and 1966, were treated with INH. At the time of reporting none had died of cancer. During the 13-year period, 665 live births were recorded. Five of the offspring were subsequently untraced, but the remainder were alive at the end of the study.

Campbell & Guilfoyle (1970) reported data from a study of 3064 male tuberculosis patients under treatment between 1961 and 1966. Among 129 of these who had received INH treatment, there were 17 cases of lung cancer observed versus 15.5 expected. This difference is not statistically significant.

A number of studies have been reported by Ferebee (1965, 1970) in which tuberculosis cases were given 5 mg INH/kg bw daily for 1 year and observed for 10 years. In 10,531 patients treated in mental institutions and 12,439 household contacts treated prophylactically there was no indication of an increase in the frequency of cancer deaths after 10 years of observation, in comparison with approximately similar numbers of controls given placebos. These trials are to be extended to 15 years.

A small percentage of cases of lupus vulgaris are known to have become malignant, and many of the patients treated with INH will also have had treatment with other agents. Although Pompe (1956) found some indication that the introduction of INH had been accompanied by an increase in the malignancy rate, subsequent studies (Michalowski & Kudejko, 1965; Nyfors, 1968; Jung, 1968, 1971) failed to confirm this.

In summary, all the studies mentioned above have failed to show any significant excess of cancer among patients treated with INH. However, as this drug was little used prior to 1952, the negative observations refer to a period of time not exceeding 15 years, which may not be long enough to establish the absence of a carcinogenic effect.

4. Comments on Data Reported and Evaluation

4.1 Animal data

Isonicotinic acid hydrazide (INH) is carcinogenic in mice after oral, subcutaneous and intraperitoneal administration. The observation of tumours in rats in only one of several oral studies is inconclusive. INH failed to produce tumours in hamsters when given orally.

4.2 Human data

Available evidence from the first 15 years of human exposure has not suggested that INH is carcinogenic in man in the doses applicable to treatment and prophylaxis of tuberculosis.

5. References

Biancifiori, C., Bucciarelli, E., Santilli, F.E. & Ribacchi, R. (1963) Carcinogenesi polmonare da idrazide dell' acido isonicotinico (INI) e suoi metaboliti in topi CBA/Cb/Se substrain. Lav. Ist. Anat. Univ. Perugia, 23, 209

Biancifiori, C. & Ribacchi, R. (1962a) Pulmonary tumours in mice induced by oral isoniazid and its metabolites. Nature (Lond.), 194, 488

Biancifiori, C. & Ribacchi, R. (1962b) The induction of pulmonary tumours in BALB/c mice by oral administration of isoniazid. In: Severi, L., ed., The Morphological Precursors of Cancer, Perugia, Division of Cancer Research, p. 63

Biancifiori, C. & Severi, L. (1966) The relation of isoniazid (INH) and allied compounds to carcinogenesis in some species of small laboratory animals: a review. Brit. J. Cancer, 20, 528

Campbell, A.H. & Guilfoyle, P. (1970) Pulmonary tuberculosis, isoniazid and cancer. Brit. J. Dis. Chest, 64, 141

Canbäck, T. (1952) A note on the estimation of isonicotinic acid hydrazide. J. Pharm. Pharmacol., 4, 407

Colvin, L.B. (1969) Metabolic fate of hydrazines and hydrazides. J. pharm. Sci., 58, 1433

Dymond, L.C. & Russell, D.W. (1970) Rapid determination of isonicotinic acid hydrazide in whole blood with 2,4,6-trinitrobenzenesulphonic acid. Clin. Chim. Acta, 27, 513

Ellard, G.A., Gammon, P.T. & Wallace, S.M. (1972) The determination of isoniazid and its metabolites acetylisoniazid, monoacetylhydrazine, diacetylhydrazine, isonicotinic acid and isonicotinylglycine in serum and urine. Biochem. J., 126, 449

Ferebee, S.H. (1965) Carcinogenesis of isoniazid in mice. Brit. med. J., ii, 1122

Ferebee, S.H. (1970) Controlled chemoprophylaxis trials in tuberculosis. A general review. Advanc. Tuberc. Res., 17, 28

Finkle, B.S., Cherry, E.J. & Taylor, D.M. (1971) A GCL based system for the detection of poisons, drugs and human metabolites encountered in forensic toxicology. J. chromat. Sci., 9, 393

Frohberg, H. (1970) Derzeitiger Stand der Arzneimittel-Toxikologie. München Med. Wschr., 112, 1532

Gelber, R., Peters, J.H., Gordon, G.R., Glazko, A.J. & Levy, L. (1971) The polymorphic acetylation of dapsone in man. Clin. pharmacol. Ther., 12, 225

Gemeinhardt, K. & Rangnick, G.F. (1953) Zur Analytik des Isonikotinsäure-hydrazids. Arzneimittel-Forsch., 3, 45

Hammond, E.C., Selikoff, I.J. & Robitzek, E.H. (1967) Isoniazid therapy in relation to later occurrence of cancer in adults and in infants. Brit. med. J., ii, 792

Hughes, H.B. (1953) On the metabolic fate of isoniazid. J. Pharmacol. exper. Ther., 109, 444

Iwainsky, H. (1965) Chemische und biochemische Untersuchungen zum Stoffwechsel des Isonicotinsäurehydrazids und seiner Derivate - ein Beitrag zum Fremdstoffproblem bei Nahrung und Ernährung.

Jones, L.D., Fairchild, D.G. & Morse, W.C. (1971) The induction of pulmonary neoplasms in mice by isonicotinic acid hydrazide. Amer. Rev. resp. Dis., 103, 612

Juhász, J., Baló, J. & Kendrey, G. (1957) Über die geschwulsterzeugende Wirkung des Isonicotinsäurehydrazid (INH). Z. Krebsforsch., 62, 188

Juhász, J., Baló, J. & Szende, B. (1963) Neue experimentelle Angaben zur geschwulsterzeugenden Wirkung des Isonikotinsäurehydrazid (INH). Z. Krebsforsch., 65, 434

Jung, H.D. (1968) 12 Jahre Bekämpfung der Hauttuberkulose im Landbezirk Neubrandenburg - eine sozialhygienische Aufgabenstellung. Beitr. Klin. Tuberk., 138, 206

Jung, H.D. (1971) Zur fraglichen Induzierung von Bronchial- und Hautkarzinomen durch INH (Isoniazid) am Beispiel der Tuberculosis luposa. Z. Erkr. Atmungsorgane, 135, 31

Kelly, M.G., O'Gara, R.W., Yancey, S.T., Gadekar, K., Botkin, C. & Oliviero, V.T. (1969) Comparative carcinogenicity of N-isopropyl-α-(2-methylhydrazino)-p-toluamide·HCl (procarbazine hydrochloride), its degradation products, other hydrazines and isonicotinic acid hydrazide. J. nat. Cancer Inst., 42, 337

Kelly, J.M. & Poet, R.B. (1952) The estimation of isonicotinic acid hydrazide (Nydrazid) in biological fluids. Amer. Rev. Tuberc., 65, 484

Kern, W., ed. (1958) Tuberkulosemittel. In: Hagers Handbuch der Pharmazeutischen Praxis, Berlin, Göttingen, Heidelberg, Springer-Verlag, Ergänzungsband II, p. 1866

Krüger-Thiemer, E. (1956) Chemie des Isoniazids. In: Freerksen, E., ed., Jahresbericht Borstel, 3, Berlin, Göttingen, Heidelberg, Springer-Verlag, p. 323

Krüger-Thiemer, E. (1957) Biochemie des Isoniazids. I. Isoniazidmetabolismus. In: Freerksen, E., ed., Jahresbericht Borstel, 4, Berlin, Göttingen, Heidelberg, Springer-Verlag, p. 299

Leuschner, F. (1954) Eine chromatographische Methode zur Bestimmung des Isonicotinsäurehydrazids. Arch. exp. Pathol. Pharmakol., 221, 323

Linnik, A.B. (1972) A study of the possible blastomogenic action of tubazid and phtivazid in the experiments on animals. Vop. Onkol., 18, 54

Loscalzo, B. (1964) Hydrazide de l'acide isonicotinique et neoplasies. Arch. int. Pharmacodyn., 152, 249

Lucchesi, M., Storniello, G. & Zubiani, M. (1967) In dagini sperimentali sulla presunta azione oncogena dell'isoniazide negli animali da laboratorio. Ann. Ist. Forlanini, 27, 62

Merck & Co. Inc. (1968) The Merck Index, 8th ed., 586

Michalowski, R. & Kudejko, T. (1965) Karzinomentstehung in mit Vitamin D_2 und Isonicotinsäurehydrasid behandeltem Lupus vulgaris. Derm. Wschr., 151, 25

Mori, K. & Yasuno, A. (1959) Preliminary note on the induction of pulmonary tumors in mice by isonicotinic acid hydrazide feeding. Gann, 50, 107

Mori, K., Yasuno, A. & Matsumoto, K. (1960) Induction of pulmonary tumors in mice with isonicotinic acid hydrazide. Gann, 51, 83

Nair, V.R. & Nair, C.G.R. (1971) The titration of isoniazid and other hydrazine derivatives with chloramine-T. Analyt. Chim. Acta, 57, 429

Nyfors, A. (1968) Lupus vulgaris, isoniazid and cancer. Scand. J. resp. Dis., 49, 264

Peacock, A. & Peacock, P.R. (1966) The results of prolonged administration of isoniazid to mice, rats and hamsters. Brit. J. Cancer, 20, 307

Peters, J.H., Miller, K.S. & Brown, P. (1965) The determination of isoniazid and its metabolites in human urine. Analyt. Biochem., 12, 379

Pompe, K. (1956) Einfluss von Isonicotinhydrazid auf die Lupuskarzinomentstehung. Derm. Wschr., 133, 105

Porcellati, G. & Preziosi, P. (1954) Trasformazione biologica dell' idrazide dell' acido isonicotinico. Enzymologia, 17, 47

Prescott, B., Kauffmann, G. & James, W.D. (1953) Rapid colorimetric method for determination of isonicotinic acid hydrazide in blood plasma. Proc. Soc. exp. Biol. (N.Y.), 84, 704

Rao, P.V.K. & Rao, G.B.B. (1971a) Vanadametric determination of isonicotinic acid hydrazide using redox indicators and application to pharmaceutical preparations. Z. analyt. Chem., 256, 360

Rao, P.V.K. & Rao, G.B.B. (1971b) A potentiometric procedure for the assay of isonicotinic acid hydrazide (isoniazid). Analyst, 96, 712

Ribacchi, R., Biancifiori, C., Milia, U., DiLeo, F.P. & Bucciarelli, E. (1963) Cancerogenesi polmonare da idrazide dell' acido isonicotinico in topi maschi BALB/c, con e senza MTV. Lav. Ist. Anat. Univ. Perugia, 23, 103

Schulte, E.H., Schloot, W. & Goedde, H.W. (1970) Zur Acetylierung von 5-Hydroxytryptamin (Serotonin) und Isoniazid durch eine N-Acetyltransferase. Hoppe-Seyler's Z. physiol. Chem., 351, 1324

Schwan, S. (1962) Isonicotinic acid hydrazide (INH) as a carcinogenic agent in mice. Second report. Pat. Pol., 13, 185

Scott, P.G.W. (1952) The detection and determination of isonicotinyl hydrazide. J. Pharm. Pharmacol., 4, 681

Severi, L. & Biancifiori, C. (1968) Hepatic carcinogenesis in CBA/Cb/Se mice and Cb/Se rats by isonicotinic acid hydrazide and hydrazine sulfate. J. nat. Cancer Inst., 41, 331

Smith, C.C. (1966) Role of nonhuman primates in predicting metabolic disposition of drugs in man. In: Proceedings of the Food and Drug Administration Conference on Nonhuman Primate Toxicology, 1966, Warrenton, Virginia, US Department of Health, Education and Welfare, p. 57

Steinberg, M.S., Cohen, S.N. & Weber, W.W. (1971) Isotope exchange studies on rabbit liver N-acetyltransferase. Biochim. biophys. Acta, 235, 89

Toida, I. (1962) Isoniazid - hydrolysing enzyme of mycobacteria. Amer. Rev. resp. Dis., 85, 720

Toth, B. & Boreisha, I. (1969) Tumorigenesis with isonicotinic acid hydrazide and urethan in the Syrian golden hamsters. Europ. J. Cancer, 5, 165

Toth, B. & Shimizu, H. (1973) Lung carcinogenesis with 1-acetyl-2-iso-nicotinoylhydrazine, the major metabolite of isoniazid. Europ. J. Cancer, 9, 285

Toth, B. & Shubik, P. (1966a) Carcinogenesis in Swiss mice by isonicotinic acid hydrazide. Cancer Res., 26, 1473

Toth, B. & Shubik, P. (1966b) Mammary tumour inhibition and lung adenoma induction by isonicotinic acid hydrazide. Science, 152, 1376

Toth, B. & Shubik, P. (1969) Lack of carcinogenic effects of isonicotinic acid hydrazide in the Syrian golden hamster. Tumori, 55, 127

Toth, B. & Toth, T. (1970) Investigation on the tumor producing effect of isonicotinic acid hydrazide in ASW/Sn mice and MRC rats. Tumori, 56, 315

US Tariff Commission (1972) Synthetic Organic Chemicals, United States Production and Sales, 1970, TC Publication 479

Vivien, J.N., Thibier, R. & Lepeuple, A. (1972) Recent studies on iso-niazid. Advanc. Tuberc. Res., 18, 148

Weinstein, H.J. & Kinosita, R. (1963) Isoniazid and pulmonary tumours in mice. Amer. Rev. resp. Dis., 88, 124

Yamamoto, R.S. & Weisburger, J.H. (1970) Failure of arginine glutamate to inhibit lung tumor formation by isoniazid and hydrazine in mice. Life Sci., 9, 285

Zamboni, V. & Fachinelli, E. (1954) Su alcuni antagonisti della isoniazide. Giorn. ital. Chemioter., 1, 638

MALEIC HYDRAZIDE[*]

1. Chemical and Physical Data

1.1 Synonyms and trade names

Chem. Abstr. No.: 123-33-1

1,2-Dihydropyridazine-3,6-dione; 1,2-dihydro-3,6-pyridazinedione; 6-hydroxy-3-(2H)pyridazinone; MAH; maleic acid hydrazide; maleic hydrazine; N,N-maleoylhydrazine; MH

Malazide; Regulox

1.2 Chemical formula and molecular weight

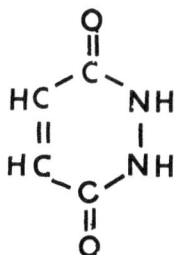

$C_4H_4N_2O_2$ Mol. wt: 112.1

1.3 Chemical and physical properties of the pure substance

(a) Description: Colourless crystalline solid

(b) Melting-point: Decomposes at about 260°C

(c) Solubility: Slightly soluble in water and hot ethanol; soluble in aqueous alkali and certain organic bases

(d) Stability: Maleic hydrazide (MH) is very stable to both acidic and basic hydrolysis (Wood, 1953)

(c) Chemical reactivity: MH is slightly acidic, may be titrated as a monobasic acid, and forms salts with alkalies and amines such as diethanolamine and triethanolamine (Schoene & Hoffmann, 1949).

---

[*] Considered by the Working Group in Lyon, June 1973.

1.4 Technical products and impurities

MH is available as a technical grade material containing 97% minimum active ingredient and less than 1% anionic wetting agent. It is also offered for sale as emulsifiable concentrates or wettable powders for agricultural uses in the form of its potassium salt or its diethanolamine salt. The products can contain small amounts of hydrazine as an impurity.

2. Production, Use, Occurrence and Analysis

Reviews concerning MH are available (Crafts, 1961; Zukel, 1957, 1963).

(a) Production and use[1]

MH was first introduced in the United States as a commercial product in 1948. It is made commercially by the condensation of maleic anhydride with dihydrazine sulphate.

Total production of the five US producers was reported to be 1.3 million kg in 1969 (US Tariff Commission, 1971) and 1.5 million kg in 1970 (US Tariff Commission, 1972). Total sales were reported to be 1.4 million kg in 1970 (US Tariff Commission, 1972), and preliminary data indicate that total sales were 1.8 million kg in 1971 (US Tariff Commission, August 1972).

US imports of MH through the principal customs districts were reported to have been 45 thousand kg in 1971 (US Tariff Commission, July 1972).

No information was found on the quantity of MH produced in countries other than the US. In 1967, one company was reported to be manufacturing MH in the Federal Republic of Germany, and one producer was reported in Italy in 1969. Seven Japanese companies were reported to be producing hydrazine derivatives in 1972, and it is probable that one or more of these companies produced MH. At least one French company was offering MH for sale in 1972, but whether this company was also manufacturing the product is not known.

---

[1] Data from Chemical Information Services, Stanford Research Institute, USA.

174

MH is used exclusively as a plant growth inhibitor and herbicide. The principal use is in the control of sucker growth on tobacco. One source estimated that maleic anhydride was applied to 30-40% of the tobacco grown in the US in 1961 (US Department of Agriculture, September 1963). A later survey of farmers indicated that 1.5 million kg of MH were used by US farmers in 1966, and that almost all of this was used on crops (US Department of Agriculture, 1970). In addition to being used on tobacco, maleic anhydride was approved in the US for use as a plant regulator on the following other crops, as of September 1968: beans, beets, citrus fruits, corn, lima beans, onions, peas, potatoes, strawberries, sugar beets and tomatoes. It was also approved for onion and garlic control in pasture land (US Department of Agriculture, 1968). Its second most important use, after that on tobacco, is believed to be on potatoes and onions to control sprouting.

No information was found on the uses of MH in countries other than the US.

(b) Occurrence

MH has not been reported to occur as such in nature. Because of its use on agricultural crops, it could be expected to appear in the treated plants and in the surface waters from treated acreage. For example, MH has been reported to be present in tobacco and tobacco smoke (Wynder & Hoffmann, 1968; Liu & Hoffmann, 1973). The US Food Additives Regulations limit the quantity of MH allowable in potato chips to 160 parts per million by weight of the finished food (US Department of Health, Education and Welfare, 1970).

(c) Analysis

Small amounts (less than 1 ppm) of MH in plant and animal tissues can be estimated photometrically (minimum of transmittance at 445 nm) after reduction with zinc and sodium hydroxide, hydrolysis, and distillation of the liberated hydrazine, followed by its reaction with p-dimethyl-aminobenzaldehyde (Wood, 1953). It may also be determined in extracts from tobacco smoke by gas chromatographic methods, using the bis(4-chlorobenzyl)-derivative (limit of detection, 1 ng) (Liu & Hoffmann, 1973).

175

3. Biological Data Relevant to the Evaluation
of Carcinogenic Risk to Man

3.1 Carcinogenicity and related studies in animals

(a) Oral administration

Mouse: Innes et al.(1969) administered MH in daily doses of
1000 mg/kg bw by stomach tube to 36 mice of each sex for 3 weeks,
beginning when the animals were 7 days old. Then, 3000 ppm were
mixed directly with the diet which was fed ad libitum for approximately 18 months. No significant increase in the incidence of tumours
was observed, in comparison with untreated controls.

Rat: Barnes et al.(1957) reported that 30 rats fed 1% MH in
the diet for 100 weeks failed to show a significant increase in the
number of tumours, in comparison with that found in 20 control animals.

(b) Subcutaneous and/or intramuscular administration

Mouse: A group of 27 mice injected s.c. with weekly doses of
500 mg/kg bw monosodium salt of MH did not show any difference in
tumour incidence as compared with that found in 20 controls (Barnes
et al., 1957).

Newborn mice: Swiss mice were injected s.c. on days 1, 7, 14
and 21 following birth with a total dose of 3 mg MH per mouse as an
aqueous solution. The MH used contained 0.4% hydrazine as impurity.
Hepatomas developed in 3/17 male survivors after 49 weeks. In a
second group, injected with MH suspended in tricaprylin at a total
dose of 55 mg per animal, 19/26 male survivors (65%) had hepatomas
at about 50 weeks. In control males, 4/48 hepatomas were seen. No
hepatomas developed in female animals (Epstein et al., 1967; Epstein
& Mantel, 1968).

Rat: Barnes et al.(1957) reported that 500 mg/kg bw monosodium
salt of MH given s.c. weekly for 100 weeks failed to show significant
differences in the induction of local and systemic tumours in 29
treated rats as compared with the incidence in 20 controls. Similar
results were obtained by Dickens & Jones (1965) in 6 rats surviving

up to 106 weeks after twice weekly s.c. injections of 2 mg MH (as the free substance, not the salt) for 65 weeks (total dose, 260 mg). No tumours were found in 6 rats surviving 45 or 106 weeks after injections of arachis oil alone.

3.2 Other relevant biological data

(a) Animals

No data on enzymatic metabolism in animals or carcinogenicity of metabolites are available. Barnes et al.(1957) established that after oral doses of 100 mg/kg bw, rabbits excrete 43 to 62% MH unchanged in the urine within 48 hours.

3.3 Observations in man

Human populations may be exposed to MH (Bishop & Schweers, 1961), e.g., by inhalation when the substance is sprayed for agricultural and horticultural purposes, and by intake with the food, but no epidemiological studies are known to the Working Group.

4. Comments on Data Reported and Evaluation[1]

4.1 Animal data

No carcinogenic effect was observed in adult mice and rats following oral or subcutaneous administration of maleic hydrazide (MH). The significance of hepatomas obtained in newborn mice cannot be assessed because of the contamination of maleic hydrazide with hydrazine.

4.2 Human data

No epidemiological data are available to the Working Group.

---

[1] See also the section "Extrapolation from animals to man" in the introduction to this volume.

5. References

Barnes, J.M., Magee, P.N., Boyland, E., Haddow, A., Passey, R.D., Bullough, W.S., Cruickshank, C.N.D., Salaman, M.H. & Williams, R.T. (1957) The non-toxicity of maleic hydrazide for mammalian tissues. Nature(Lond.), 180, 62

Bishop, J.C. & Schweers, V.H. (1961) Sprout inhibition of fall-grown potatoes by airplane applications of maleic hydrazide. Amer. Potato J., 38, 377

Crafts, A.S. (1961) The Chemistry and Mode of Action of Herbicides, New York, Wiley Interscience, p. 186

Dickens, F. & Jones, H.E.H. (1965) Further studies on the carcinogenic action of certain lactones and related substances in the rat and mouse. Brit. J. Cancer, 19, 392

Epstein, S.S., Andrea, J., Jaffe, H., Joshi, S., Falk, H. & Mantel, N. (1967) Carcinogenicity of the herbicide maleic hydrazide. Nature (Lond.), 215, 1388

Epstein, S.S. & Mantel, N. (1968) Hepatocarcinogenicity of the herbicide maleic hydrazide following parenteral administration to infant Swiss mice. Int. J. Cancer, 3, 325

Innes, J.R.M., Ulland, B.M., Valerio, M.G., Petrucelli, L., Fishbein, L., Hart, E.R., Pallotta, A.J., Bates, R.R., Falk, H.L., Gart, J.J., Klein, M., Mitchell, I. & Peters, J. (1969) Bioassay of pesticides and industrial chemicals for tumorigenicity in mice: a preliminary note. J. nat. Cancer Inst., 42, 1101

Liu, Y.-Y. & Hoffmann, D. (1973) Quantitative determination of maleic hydrazide in cigarette smoke. Analyt. Chem. (in press)

Schoene, D.L. & Hoffmann, O.L. (1949) Maleic hydrazide, a unique growth regulant. Science, 109, 588

US Department of Agriculture (September 1963) The Pesticide Situation for 1962-1963, Agricultural Stabilization and Conservation Service

US Department of Agriculture (1968) Summary of Registered Agricultural Pesticide Chemical Uses, Pesticides Regulation Division, Agricultural Research Service, Vol. I, I-D-12.1 and I-D-12.2

US Department of Agriculture (1970) Quantities of Pesticides Used by Farmers in 1966, Economic Research Service, Agricultural Economic Report No. 179

178

US Department of Health, Education and Welfare, Food and Drug Administration (1970) Food Additives Regulations Under the Federal Food, Drug and Cosmetics Act, Part 121, Chapter I, Title 21, Code of Federal Regulations, Paragraph 121.1006

US Tariff Commission (1971) Synthetic Organic Chemicals, United States Production and Sales, 1969, TC Publication 412

US Tariff Commission (July 1972) Imports of Benzenoid Chemicals and Products, 1971, TC Publication 496

US Tariff Commission (August 1972) Synthetic Organic Chemicals, United States Production and Sales of Pesticides and Related Products, 1971 Preliminary

US Tariff Commission (1972) Synthetic Organic Chemicals, United States Production and Sales, 1970, TC Publication 479

Wood, P.R. (1953) Determination of maleic hydrazide residues in plant and animal tissue. Analyt. Chem., 25, 1879

Wynder, E.L. & Hoffmann, D. (1968) Experimental tobacco carcinogenesis. Science, 162, 862

Zukel, J.W. (1957, 1963) A Literature Summary on Maleic Hydrazide 1949-1957 and 1957-1963, Naugatuck, Connecticut, United States Rubber Company

N-NITROSO COMPOUNDS

N-METHYL-N'-NITRO-N-NITROSOGUANIDINE*

1. Chemical and Physical Data

1.1 Synonyms and trade names

Chem. Abstr. No.: 70-25-7

1-Methyl-3-nitro-1-nitrosoguanidine; N-methyl-N-nitroso-N'-nitroguanidine; MNG; MNNG; NG

1.2 Chemical formula and molecular weight

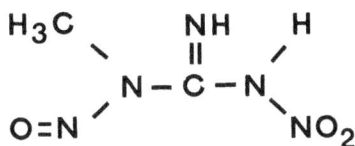

$$H_3C \diagdown \quad NH \quad H$$
$$ N - C - N \diagup$$
$$O=N \diagup \diagdown NO_2$$

$C_2H_5N_5O_3$ Mol. wt: 147.1

1.3 Chemical and physical properties of the pure substance

(a) Description: Pale yellow to pink crystals

(b) Melting-point: 118-123.5°C (with decomposition)

(c) Absorption spectroscopy: λ_{max} 275 nm; log ε 4.26

$ \lambda_{max}$ 306 nm; log ε 3.18

$ \lambda_{max}$ 402 nm; log ε 2.29 (in methanol)

(d) Solubility: Only slightly soluble in water (less than 0.5%); soluble in polar organic solvents, however, this is often accompanied by decomposition.

(e) Stability: The pure compound is sensitive to light, changing to orange and green colours. Degradation products arising from prolonged or inadequate storage include N-methyl-

---

* Considered by the Working Group in Lyon, June 1973.

N'-nitroguanidine, N-nitroguanidine, nitrocyanamide and guanidine (Sugimura et al., 1969). N-Methyl-N'-nitro-n-nitroso-guanidine (MNNG) is much more stable than comparable alkyl-nitrosoureas and alkylnitrosourethanes (Druckrey et al., 1967). Thus, at room temperature, the half-life at pH 8 is about 200 hours (Druckrey et al., 1966). At pH 7.0 (phosphate-buffer) and $37^{\circ}C$, the half-life is 170 hours (Schaper, 1970). Sugimura et al. (1969) have shown that tap water decomposes MNNG much more rapidly than does deionized water. After 50 hours in deionized water, approximately 10% was lost, while in tap water more than 50% was lost. In contrast to these results, much shorter half-lives have been reported (5 hours at pH 7, and 0.75 hours at pH 8). These differences may be accounted for by the use of different buffers (McCalla et al., 1968). Decomposition kinetic data have been presented by Haga et al. (1972).

(f) Chemical reactivity: At acid pH, MNNG slowly releases nitrous acid to give N-methyl-N'-nitroguanidine (McKay & Wright, 1947). It is converted by concentrated aqueous alkali hydroxide to diazomethane (McKay, 1948). Reactions with several nucleophiles are known, especially with amines (McKay, 1949) and thiols (Schulz & McCalla, 1969).

1.4 Technical products and impurities

No data on technical products and impurities are available to the Working Group.

2. Production, Use, Occurrence and Analysis

A monograph on the analysis and formation of N-nitroso compounds has been published (Bogovski et al., 1972).

(a) Production and use[1]

MNNG can be produced by the nitrosation of N-methyl-N'-nitroguanidine.

---

[1] Data from Chemical Information Services, Stanford Research Institute, USA.

No evidence was found that MNNG is produced commercially except as a research chemical or that there is any human exposure except for that of researchers engaged in the study of its biological properties.

In the late 1940's and the 1950's, it may have found considerable use in the laboratory preparation of diazomethane solutions, since it was apparently produced commercially at that time. In some countries it has been displaced by N-nitrosomethyl-p-toluenesulphonamide as a reagent for making diazomethane.

(b) Occurrence

No data are available to the Working Group. As with other N-nitroso compounds it can be formed (Mirvish, 1971) from its precursor, N-methyl-N'-nitroguanidine, by reaction with nitrosating agents such as nitrite in environmental media and also in the gastrointestinal tract. While nitrite and other nitrosating agents are almost ubiquitous in the human environment, no data are available as to an occurrence of N-methyl-N'-nitroguanidine in the environment.

(c) Analysis

Colorimetric determination by denitrosation with dilute aqueous acids has been described (Schaper, 1970; Preussmann & Eisenbrand, 1972). A comparison of several methods of detection has also been published (Preussmann & Schaper-Druckrey, 1972).

3. Biological Data Relevant to the Evaluation of Carcinogenic Risk to Man

3.1 Carcinogenicity studies in animals

(a) Oral administration

Mouse: A single dose of 125 mg/kg bw MNNG given by intragastric intubation as a suspension in 30% aqueous alcohol to 6 male C3H mice induced 3 squamous cell carcinomas and 1 papilloma of the stomach in 3 mice within 11 to 21 months. Adenomas in the small bowel and liver were found in 1 other mouse; the spontaneous incidence of gastro-intestinal tumours is considered extremely rare in this species (Schoental & Bensted, 1969). Continuous administration of 50 mg/l in the drinking-water for 10 months to dd/I strain mice did not induce gastric tumours but gave rise to leiomyosarcomas in the walls of gastric cysts, neonatally grafted into the s.c. tissues in 5/26 mice (Matsuyama et al., 1970).

Rat: Continuous administration to male Wistar rats of MNNG dissolved in the drinking-water at concentrations of 33, 83 and 167 mg/1, for periods of 6 to 12 months, induced malignant tumours of the glandular stomach in high regularity, with a tumour incidence of 70% and above. The malignancies were found mainly in the pylorus and in the antrum region. Pathological diagnosis classified the tumours as adenomas and adenocarcinomas with a few leiomyosarcomas and signet-ring cell carcinomas. Additional malignant tumours, especially at high MNNG concentrations, were observed in the duodenum, jejunum and mesentery, together with papillomas in the forestomach and liver tumours (Sugimura & Fujimura, 1967; Takayama et al., 1969; Sugimura et al., 1969, 1970; Fujimura et al., 1970b; Bralow et al., 1970).

Sequential morphological changes during MNNG carcinogenesis in the glandular stomach in animals killed periodically from week 3 to week 60 have been described (Takayama et al., 1969; Saito et al., 1970). Stomach ulcers, produced by stress (immobilization and immersion in water at 23°C), seem to promote the induction of glandular stomach tumours (Takahashi et al., 1970). The random-bred Wistar rat seems to be especially sensitive to tumour induction in glandular stomach by administration of the carcinogen in the drinking-water. A 73% incidence of adenocarcinomas was obtained at the 83 mg/1 dose within 1 year, while under the same conditions adenocarcinomas in this organ were seen in none of the inbred Buffalo-Mai rats (only 2 sarcomas were seen) and in only 10% of the Wistar-Mai-Furst rats. Variations in the dose and frequency of MNNG administration did not overcome this strain resistance (Bralow et al., 1973).

A series of 1 to 5 gastric intubations at irregular intervals within 10 months of 50-100 mg/kg bw to 6 male and 8 female rats produced squamous papillomas and squamous cell carcinomas of the fore-stomach and tumours in the glandular stomach, liver and the

186

peritoneum within 1 to 2 years (Schoental, 1966; Craddock, 1968a; Schoental & Bensted, 1969).

A single dose of 50 or 250 mg/kg bw given in saline or 250 mg/kg bw given in olive oil to groups of 11 or 12 ACI rats resulted in tumour incidences of 10-45% and 66-70%, respectively. An effective number of 51 animals surviving longer than 300 days was evaluated. Malignant tumours in the glandular and forestomach were seen only at the higher dose level, while papillomas of the forestomach occurred in both groups (Hirono & Shibuya, 1972).

Hamster: Continuous treatment of 26 male golden hamsters with a concentration of 83 mg/1 MNNG in the drinking-water produced malignancies within 6 to 10 months. Twenty fibrosarcomas of the pyloric region of the glandular stomach, originating from the submucosa, were found. In animals dying after 260 days 3 adenocarcinomas and 2 fibrosarcomas were found in the duodenum (Fujimura et al., 1970a).

Two groups of 20 male hamsters were administered MNNG in the drinking-water (83 mg/1) for periods of 4 and 7 months, respectively. All animals developed tumours, especially adenocarcinomas and sarcomas of the glandular stomach. Additional malignant tumours in the oesophagus, forestomach, duodenum, intestine and adrenal gland were observed. Limited administration for 4, 6 and 8 months, respectively, of 50 mg/1 to 3 groups of 20 male golden hamsters produced tumours in almost all animals; more adenocarcinomas than fibrosarcomas were observed in the glandular stomach (Sugimura et al., 1969).

Rabbit: Administration in the drinking-water of 167 mg/1 MNNG to 10 male rabbits for 15 months produced squamous cell carcinomas in the tracheobronchial region in 3 animals living longer than 15 months. No tumours were found in the glandular stomach, but abnormal epithelial growth was observed. An adenocarcinoma of the duodenum was also found. The tracheobronchial tumours may have arisen from aspiration of the MNNG solution during drinking (Sugimura et al., 1969).

<u>Dog</u>: Four mongrel dogs received 167 mg/l MNNG in the drinking-water for 1 month, and then a reduced concentration of 87 mg/l for 14 months. Treatment was stopped at this time (i.e., at 463 days), and the animals were observed until death. All dogs developed adenocarcinomas of the stomach. Tumours were localized mainly in the cardiac portion and in the antrum. Furthermore, spindle cell sarcomas and haemangioblastomas were found in the small intestine. The latent period of tumour production was 518 to 1045 days (Sugimura et al., 1971; Shimosato et al., 1971).

(b) <u>Subcutaneous and/or intramuscular administration</u>

<u>Rat</u>: In 6/6 and 8/10 BDII rats given 5 s.c. injections of 90 or 45 mg MNNG/kg bw/week in oil, respectively, fibrosarcomas and polymorphic sarcomas developed at the site of injection after 180 to 360 days (Druckrey et al., 1966). Weekly s.c. injections of 10 or 25 mg MNNG in a 0.5% aqueous solution to 2 groups of 12 Wistar rats gave rise mainly to local fibrosarcomas and some rhabdomyosarcomas in 33% and 50% of the animals, respectively, within 300 to 360 and 170 to 333 days (Sugimura et al., 1966).

<u>Newborn rats</u>: Out of 154 ACI x Sprague-Dawley rats receiving a single s.c. dose of 2, 10 or 100 µg MNNG/rat at birth, tumours developed in 15/46 rats surviving for longer than 1 year. The tumours were mainly adenocarcinomas and fibro- and myosarcomas of the small intestine. Other sites in which malignancies were found were the stomach, liver, peritoneum, uterus and ovary. The tumour incidences were 2/10, 1/12 and 12/24 at the 3 levels, respectively (Takayama et al., 1972).

(c) <u>Skin application</u>

<u>Mouse</u>: Treatment of the skin 3 times weekly with 0.05 ml of 0.15%, 0.3% and 0.5% solutions of MNNG in acetone for 5 months induced 6 papillomas, 17 fibrosarcomas, 8 mixed tumours (composed of both squamous cell carcinomas and fibrosarcomas) and 7 squamous cell carcinomas at the site of application in 28 ICR mice. No distant tumours were seen. No local tumours were found in the controls

188

(Takayama et al., 1971).

(d) Intraperitoneal administration

Mouse: In 12 CFW and 12 C3H male mice given 2 to 3 doses of approximately 100 mg/kg bw MNNG within 6 to 10 months, 4 benign and malignant tumours of the caecum, ileum and jejunum and 1 adrenal cortical tumour were observed in animals dying after 13 to 16 months (Schoental & Bensted, 1969).

Rat: In 12 male rats a single dose of approximately 60 mg/kg bw MNNG followed by 2 to 3 doses of 10-25 mg/kg bw within 11 months led to the development of 5 malignant tumours in the stomach, jejunum and caecum after 12 to 19 months (Schoental & Bensted, 1969). In suckling ACI x Sprague-Dawley rats a single dose of 600 µg MNNG/rat resulted in papillomas, carcinomas and sarcomas of the stomach and small intestine and a few tumours at other sites in 53% of 38 animals surviving longer than 1 year (Takayama et al., 1972).

(e) Other experimental systems

Rectal administration: Daily rectal infusion for 32 days of 0.5 ml of a 0.25% aqueous solution (total dose, 40 mg/rat) of MNNG induced one or more adenomatous polyps and polypoid carcinomas in the colon and rectum in 7/9 rats living longer than 30 weeks (Narisawa et al., 1971, 1972). So et al. (1973) confirmed these results.

3.2 Other relevant biological data

After oral administration of MNNG about 90% is excreted in the urine, mostly as N-methyl-N'-nitro-guanidine, in the first 9 hours (Sugimura et al., 1972). There is evidence that denitrosation of MNNG is effected by enzymes occurring in the stomach, liver and kidney; these have been partially purified (Kawachi et al., 1970; Sugimura et al., 1972).

In vitro and in vivo, MNNG causes methylation of nucleic acids, forming mainly 7-methylguanine with smaller amounts of 3-methyladenine, 1-methyladenine, 3-methylcytosine and O^6-methylguanine (Lawley, 1968; Lawley & Thatcher, 1970; Lawley & Shah, 1972; Craddock, 1968b, 1969; McCalla, 1968; Sugimura et al., 1968; Singer et al., 1968; Süssmuth et

189

al., 1972). MNNG can modify proteins by transferring its nitroguanidine residue, e.g., by converting lysine into nitrohomoarginine and cysteine into cystine (Lawley & Thatcher, 1970; McCalla & Reuvers, 1968; Sugimura et al., 1968; Schulz & McCalla, 1969). Cytochrome c thus modified no longer acts as an electron acceptor (Nagao et al., 1971). Histones from ascites tumour cells contained nitrohomoarginine after MNNG treatment (Nagao et al., 1969).

In vitro transformation: Thymus and lung cells from suckling rats showed malignant transformation after MNNG treatment in culture. Sub-cutaneous implantation of treated cells produced sarcomas (Takaki et al., 1969; Takii et al., 1971). Transformed and transplantable cells were also obtained in cultured hamster lung cells (Inui et al., 1972). Liver cells originating from rats have been transformed in vitro as shown by the development of carcinosarcoma following back transplantation of the treated cells into syngeneic rats (Montesano et al., 1973).

3.3 Observations in man

No data are available to the Working Group.

4. Comments on Data Reported and Evaluation[1]

4.1 Animal data

N-methyl-N'-nitro-N-nitrosoguanidine (MNNG) is carcinogenic in all species tested: mouse, rat, hamster, rabbit and dog. It has a predominantly local carcinogenic effect following administration by oral and other routes. It is carcinogenic in single-dose experiments.

4.2 Human data

No epidemiological data are available to the Working Group.

---

[1] See also the section "Extrapolation from animals to man" in the introduction to this volume.

5. References

Bogovski, P., Preussmann, R. & Walker, E.A., eds (1972) N-Nitroso
   Compounds, Analysis and Formation, Lyon, IARC Scientific Publications,
   3

Bralow, S.P., Gruenstein, M. & Meranze, D.R.(1973) Host resistance to
   gastric adenocarcinomatosis in three strains of rats ingesting N-
   methyl-N'-nitro-N-nitrosoguanidine. Oncology, 27, 168

Bralow, S.P., Gruenstein, M., Meranze, D.R., Bonakdarpour, A. & Shimkin,
   M.B. (1970) Adenocarcinoma of glandular stomach and duodenum in
   Wistar rats ingesting N-methyl-N'-nitro-N-nitrosoguanidine, histo-
   pathology and associated secretory changes. Cancer Res., 30, 1215

Craddock, V.M. (1968a) The effect of N'-nitro-N-nitroso-N-methylguanidine
   on the liver after administration to the rat. Experientia (Basel),
   24, 1148

Craddock, V.M. (1968b) The reaction of N-methyl-N'-nitro-N-nitrosoguani-
   dine with deoxyriboncucleic acid. Biochem. J., 106, 921

Craddock, V.M. (1969) Study of the methylation and lack of deamination of
   deoxyribonucleic acid by N-methyl-N'-nitro-N-nitrosoguanidine.
   Biochem. J., 111, 615

Druckrey, H., Preussmann, R., Ivankovic, S. & Schmähl, D. (1967) Organotrope carcinogene Wirkungen bei 65 verschiedenen N-Nitroso-
   Verbindungen an BD-Ratten. Z. Krebsforsch., 69, 103

Druckrey, H., Preussmann, R., Ivankovic, S., So, B.T., Schmidt, C.H. &
   Bücheler, J. (1966) Zur Erzeugung subcutaner Sarkome an Ratten.
   Carcinogene Wirkung von Hydrazodicarbonsäure-bis-(methyl-nitrosamid),
   N-Nitroso-N-n-butyl-harnstoff, N-Methyl-N-nitroso-nitroguanidin und
   N-Nitroso-imidazolidon. Z. Krebsforsch., 68, 87

Fujimura, S., Kogure, K., Oboshi, S. & Sugimura, T. (1970a) Production of
   tumors in glandular stomach of hamsters by N-methyl-N'-nitro-N-nitro-
   soguanidine. Cancer Res., 30, 1444

Fujimura, S., Kogure, K., Sugimura, T. & Takayama, S. (1970b) The effect
   of limited administration of N-methyl-N'-nitro-N-nitrosoguanidine on
   the induction of stomach cancer in rats. Cancer Res., 30, 842

Haga, J.J., Russell, B.R. & Chapel, J.F. (1972) The kinetics of decomposition of N-alkyl derivatives of nitrosoguanidine. Cancer Res., 32,
   2085

Hirono, I. & Shibuya, C. (1972) Induction of stomach cancer by a single dose of N-methyl-N -nitro-N-nitrosoguanidine through a stomach tube. In: Nakahara, W., Takayama, S., Sugimura, T. & Odashima, S., eds., Topics in Chemical Carcinogenesis, Tokyo, University of Tokyo Press, p. 121

Inui, N., Takayama, S. & Sugimura, T. (1972) Neoplastic transformation and chromosomal aberrations induced by N-methyl-N'-nitro-N-nitroso-guanidine in hamster lung cells in tissue culture. J. nat. Cancer Inst., 48, 1409

Kawachi, T., Kogure, K., Kamijo, Y. & Sugimura, T. (1970) The metabolism of N-methyl-N'-nitro-N-nitrosoguanidine in rats. Biochim. biophys. Acta, 222, 409

Lawley, P.D. (1968) Methylation of DNA by N-methyl-N-nitrosourethane and N-methyl-N'-nitro-N-nitrosoguanidine. Nature (Lond.), 218, 580

Lawley, P.D. & Shah, S.A. (1972) Methylation of ribonucleic acid by the carcinogens dimethyl sulphate, N-methyl-N-nitrosourea and N-methyl-N'-nitro-N-nitrosoguanidine. Comparisons of chemical analyses at at the nucleoside and base levels. Biochem. J., 128, 117

Lawley, P.D. & Thatcher, C.J. (1970) Methylation of deoxyribonucleic acid in cultured mammalian cells by N-methyl-N'-nitro-N-nitrosoguanidine. The influence of cellular thiol concentrations on the extent of methylation and the 6-oxygen atom of guanine as a site of methylation. Biochem. J., 116, 693

Matsuyama, M., Suzuki, H. & Nakamura, T. (1970) Leiomyosarcomas induced by oral administration of N-methyl-N'-nitro-N-nitrosoguanidine in gastric cysts grafted in subcutaneous tissue of mice. Gann, 61, 523

McCalla, D.R. (1968) Reaction of N-methyl-N'-nitro-N-nitrosoguanidine and N-methyl-N-nitroso-p-toluenesulfonamide with DNA in vitro. Biochim. biophys. Acta, 155, 114

McCalla, D.R. & Reuvers, A. (1968) Reaction of N-methyl-N'-nitro-N-nitrosoguanidine with protein: formation of nitroguanido derivatives. Canad. J. Biochem., 46, 1411

McCalla, D.R., Reuvers, A. & Kitai, R. (1968) Inactivation of biologically active N-methyl-N-nitroso compounds in aqueous solution: effect of various conditions of pH and illumination. Canad. J. Biochem., 46, 807

McKay, A.F. (1948) A new method of preparation of diazomethane. J. amer. Chem. Soc., 70, 1974

McKay, A.F. (1949) The preparation of N-substituted-N'-nitroguanidines by the reaction of primary amines with N-alkyl-N-nitroso-N'-nitroguani-dines. J. amer. Chem. Soc., 71, 1968

McKay, A.F. & Wright, G.F. (1947) Preparation and properties of N-methyl-N-nitroso-N'-nitroguanidine. J. amer. Chem. Soc., 69, 3028

Mirvish, S.S. (1971) Kinetics of nitrosamide formation from alkylurea, N-alkylurethans and alkylguanidines: possible implications for the etiology of human gastric cancer. J. nat. Cancer Inst., 46, 1183

Montesano, R., Saint-Vincent, L. & Tomatis, L. (1973) Malignant transformation in vitro of rat liver cells by dimethylnitrosamine and N-methyl-N'-nitro-N-nitrosoguanidine. Brit. J. Cancer, 28, 215

Nagao, M., Hosoi, H. & Sugimura, T. (1971) Modification of cytochrome c with N-methyl-N'-nitro-N-nitrosoguanidine. Biochim. biophys. Acta, 237, 369

Nagao, M., Yokoshima, T., Hosoi, H. & Sugimura, T. (1969) Interaction of N-methyl-N'-nitro-N-nitrosoguanidine with ascites hepatoma cells in vitro. Biochim. biophys. Acta, 192, 191

Narisawa, T., Nakano, H., Hayakawa, M., Sato, T. & Sakuma, A. (1972) Tumors of the colon and rectum induced by N-methyl-N'-nitro-N-nitrosoguanidine. In: Nakahara, W., Takayama, S., Sugimura, T. & Odashima, S., eds., Topics in Chemical Carcinogenesis, Tokyo, University of Tokyo Press, p. 145

Narisawa, T., Sato, T., Hayakawa, M., Sakuma, A. & Nakano, H. (1971) Carcinoma of the colon and rectum of rats by rectal infusion of N-methyl-N'-nitro-N-nitrosoguanidine. Gann, 62, 231

Preussmann, R. & Eisenbrand, G. (1972) Problems and recent results in the analytical determination of N-nitroso compounds. In: Nakahara, W., Takayama, S., Sugimura, T. & Odashima, S., eds., Topics in Chemical Carcinogenesis, Tokyo, University of Tokyo Press, p. 323

Preussmann, R. & Schaper-Druckrey, F. (1972) Investigation of a colorimetric procedure for determination of nitrosamides and comparison with other methods. In: Bogovski, P., Preussmann, R. & Walker, E.A., eds., N-Nitroso Compounds, Analyses and Formation, Lyon, IARC Scientific Publications, 3, p. 81

Saito, T., Inokuchi, K., Takayama, S. & Sugimura, T. (1970) Sequential morphological changes in N-methyl-N'-nitro-N-nitrosoguanidine carcinogenesis in the glandular stomach of rats. J. nat. Cancer Inst., 44, 769

Schaper, F. (1970) Zur Analytik von N-Alkyl-N-Nitrosamiden. (Thesis, University of Freiburg/Br.)

Schoental, R. (1966) Carcinogenic activity of N-methyl-N-nitroso-N'-nitroguanidine. Nature (Lond.), 209, 726

Schoental, R. & Bensted, J.P.M. (1969) Gastro-intestinal tumours in rats and mice following various routes of administration of N-methyl-N-nitroso-N'-nitroguanidine and N-ethyl-N-nitroso-N'-nitroguanidine. Brit. J. Cancer, 23, 757

Schulz, U. & McCalla, D.R. (1969) Reactions of cysteine with N-methyl-N-nitroso-p-toluenesulfonamide and N-methyl-N'-nitro-N-nitrosoguanidine. Canad. J. Chem., 47, 2021

Shimosato, Y., Tanaka, N., Kogure, K., Fujimura, S., Kawachi, T. & Sugimura, T. (1971) Histopathology of tumors of canine alimentary tract produced by N-methyl-N'-nitro-N-nitrosoguanidine, with particular reference to gastric carcinomas. J. nat. Cancer Inst., 47, 1053

Singer, B., Fraenkel-Conrat, H., Greenberg, J. & Michelson, A.M. (1968) Reaction of nitrosoguanidine (N-methyl-N'-nitro-N-nitrosoguanidine) with tobacco mosaic virus and its RNA. Science, 160, 1235

So, B.T., Magadia, N.E. & Wynder, E.L. (1973) Induction of carcinomas of the colon and rectum in rats by intrarectal instillation of N-methyl-N'-nitro-N-nitrosoguanidine. J. nat. Cancer Inst., 50, 927

Sugimura, T. & Fujimura, S. (1967) Tumour production in glandular stomach of rat by N-methyl-N'-nitro-N-nitroso-guanidine. Nature (Lond.), 216, 943

Sugimura, T., Fujimura, S. & Baba, T. (1970) Tumor production in the glandular stomach and alimentary tract of the rat by N-methyl-N'-nitro-N-nitrosoguanidine. Cancer Res., 30, 455

Sugimura, T., Fujimura, S., Kogure, K., Baba, T., Saito, T., Nagao, M., Hosoi, H., Shimosato, Y. & Yokoshima, T. (1969) Production of adeno-carcinomas in glandular stomach of experimental animals by N-methyl-N'-nitro-N-nitrosoguanidine. Gann Monogr., 8, 157

Sugimura, T., Fujimura, S., Nagao, M., Yokoshima, T. & Hasegawa, M. (1968) Reaction of N-methyl-N'-nitro-N-nitrosoguanidine with protein. Biochim. biophys. Acta, 170, 427

Sugimura, T., Kawachi, T., Kogure, K., Nagao, M., Tanaka, N., Fujimura, S., Takayama, S., Shimosato, Y., Noguchi, M., Kuwabara, N. & Yamada, T. (1972) Induction of stomach cancer by N-methyl-N'-nitro-N-nitroso-guanidine: Experiments on dogs as clinical models and the metabolism of this carcinogen. In: Nakahara, W., Takayama, S., Sugimura, T. & Odashima, S., eds., Topics in Chemical Carcinogenesis, Tokyo, University of Tokyo Press, p. 105

Sugimura, T., Nagao, M. & Okada, Y. (1966) Carcinogenic action of N-methyl-N'-nitro-N-nitrosoguanidine. Nature (Lond.), 210, 962

Sugimura, T., Tanaka, N., Kawachi, T., Kogure, K., Fujimura, S. & Shimosato, Y. (1971) Production of stomach cancer in dogs by N-methyl-N'-nitro-N-nitrosoguanidine. Gann, 62, 67

Süssmuth, R., Haerlin, R. & Lingens, F. (1972) The mode of action of N-methyl-N'-nitro-N-nitrosoguanidine in mutagenesis. VII. The transfer of the methyl group of N-methyl-N'-nitro-N-nitrosoguanidine. Biochim. biophys. Acta, 269, 276

Takahashi, A., Onoda, K.I., Kawashima, K., Kato, R., Omori, Y. & Ishidate, M. (1970) Effect of stress on formation of stomach tumor in rats by N-methyl-N'-nitro-N-nitrosoguanidine. Gann, 61, 295

Takaki, R., Takii, M. & Ikegami, T. (1969) Preliminary studies of the in vitro carcinogenesis of rat thymus cells by N-methyl-N'-nitro-N-nitrosoguanidine. Gann, 60, 661

Takayama, S., Kuwabara, N., Azama, Y. & Sugimura, T. (1971) Skin tumors in mice painted with N-methyl-N'-nitro-N-nitrosoguanidine and N-ethyl-N'-nitro-N-nitrosoguanidine. J. nat. Cancer Inst., 46, 973

Takayama, S., Kuwabara, N., Nemoto, N. & Azama, Y. (1972) Carcinogenesis in newborn and suckling rats induced by N-methyl-N'-nitro-N-nitroso-guanidine. In: Nakahara, W., Takayama, S., Sugimura, T. & Odashima, S., eds., Topics in Chemical Carcinogenesis, Tokyo, University of Tokyo Press, p. 133

Takayama, S., Saito, T., Fujimura, S. & Sugimura, T. (1969) Histological findings of gastric tumors induced by N-methyl-N'-nitro-N-nitroso-guanidine in rats. Gann Monogr., 8, 197

Takii, M., Takaki, R. & Okada, N. (1971) Carcinogenesis in tissue culture. XVI. Malignant transformation of rat cells by N-methyl-N'-nitro-N-nitrosoguanidine. Jap. J. exp. Med., 41, 563

N-NITROSO-DI-n-BUTYLAMINE*

1. Chemical and Physical Data

1.1 Synonyms and trade names

Chem. Abstr. No.: 924-16-3

Dibutylnitrosamine; N,N-di-n-butylnitrosamine; N,N-dibutylnitroso-amine; DBN; DBNA

1.2 Chemical formula and molecular weight

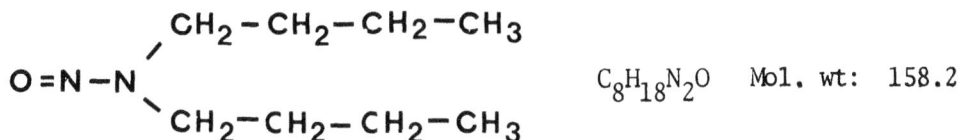

$$O = N - N \begin{array}{c} CH_2 - CH_2 - CH_2 - CH_3 \\ \\ CH_2 - CH_2 - CH_2 - CH_3 \end{array}$$

$C_8H_{18}N_2O$ Mol. wt: 158.2

1.3 Chemical and physical properties of the pure substance

(a) Description: A pale yellow liquid with a characteristic odour

(b) Boiling-point: 234-237°C (113-115°C at 14 mm)

(c) Density: D_4^{20} 0.9009

(d) Refractive index: n_D^{20}

(e) Absorption spectroscopy: λ_{max} 233 nm; log ε 3.85

λ_{max} 347 nm; log ε 1.95 (in water)

(f) Solubility: 0.12% dissolves in water at room temperature; miscible with hexane and dichloromethane, and probably with many other organic solvents

(g) Volatility: Steam volatile

(h) Stability: Stable at room temperature for more than 14 days in aqueous solutions at neutral and alkaline pH in the absence of light (Druckrey et al., 1967); slightly less stable at the pH

---

* Considered by the Working Group in Lyon, June 1973.

of strong acids. It is light sensitive, especially to ultra-violet light, so should be stored in dark bottles.

(i) Chemical reactivity: N-Nitroso-di-n-butylamine (DBN) can be oxidized by strong oxidants to the corresponding nitramine (Sen, 1970; Althorpe et al., 1970) or reduced by various reducing agents to the corresponding hydrazine and/or amine. In the presence of HBr under anhydrous conditions, it splits readily to the amine and nitrosyl bromide (Eisenbrand & Preussmann, 1970). DBN can be reduced electrochemically to the amine and nitrite in alkaline solution (Alliston et al., 1972). It is photochemically reactive (Burgess & Lavanish, 1964; Fridman et al., 1971).

1.4 Technical products and impurities

No data are available to the Working Group.

2. Production, Use, Occurrence and Analysis

A monograph on the analysis and formation of nitrosamines has been published (Bogovski et al., 1972).

(a) Production and use[1]

DBN can be produced by nitrosation of di-n-butylamine with nitrites. No evidence was found that DBN is produced commercially except as a research chemical or that there is any human exposure except for that of researchers engaged in the study of its biological properties. DBN has been used in the synthesis of di-n-butylhydrazine (Joffe, 1958; Derr, 1960). It has also been tested for fungistatic activity (Zsolnai, 1961).

(b) Occurrence

At present, no data are available on the occurrence of DBN in the human environment.

----

[1] Data from Chemical Information Services, Stanford Research Institute, USA.

As with other nitrosamines, DBN may be formed from ingested secondary or tertiary amines and quaternary ammonium salts containing the dibutyl-amino group $(CH_3CH_2CH_2CH_2)N-$ by reaction with nitrosating agents such as nitrite in the stomach or during cooking processes (Sander, 1967; Lijinsky & Epstein, 1970). Also, such formation is possible from appropriate amine precursors (Ender & Čeh, 1971) and nitrate in the presence of nitrato-reducing bacteria (Hawksworth & Hill, 1971). Thus the possibility cannot be excluded that DBN may be formed in the human stomach by the reaction of nitrites (ingested as such or produced from ingested nitrates) with compounds containing dibutylamino groups which have been intentionally or accidentally ingested. Such compounds could include the following:-
(i) bis(ethyl(2-hydroxyethyl) dimethylammonium)sulphate bis(dibutyl-carbamate) - dibutoline sulphate (an anticholinergic recently discontinued by the US manufacturer); (ii) α-dibutylamino-α-(p-methoxyphenyl) acetamide - dibutamide (an antispasmodic which appears to have been discontinued); (iii) 3-dibutylaminomethyl-4,5,6-trihydroxy-1-isobenzo-furanone - anthallan (an antihistiminic which appears to have been discontinued); (iv) 3-dibutylaminopropyl-p-aminobenzoate sulphate - butacaine sulphate (a topical anaesthetic sometimes used on the eye, mouth, nose and throat); (v) dibutylammonium tetrafluoroborate (a surface treating agent used on aluminium products); (vi) N,N-dibutyl-p-chloro-benzenesulphonamide (a synergist used with DDT in applications to barns and other agricultural premises); and (vii) N,N-dibutyl-4-hexyloxy-1-naphthamidine hydrochloride - butamidine hydrochloride (used as an anthelmintic in the past, but which appears to have been discontinued).

(c) Analysis

DBN may be detected by colorimetric, chromatographic and polaro-graphic procedures (UICC, 1970). Chromatography employing nitrogen selective detectors has also been used for screening purposes (Fiddler et al., 1971; Rhoades & Johnson, 1970). However, these methods when used alone are not sufficiently reliable for the confirmation of trace amounts in a complex matrix such as occurs in an extract from foodstuffs (Preussmann & Eisenbrand, 1972). Confirmation by combined gas chromato-graphy and mass spectrometry is recommended (Heyns & Röper, 1971; Telling et al., 1971; Gough & Webb, 1972, 1973). Detection limits of 0.2-1 ppb, based on the amount present prior to extraction or distillation, have been obtained routinely. The formation of derivatives of DBN, such as the nitramines or fluorinated amides, followed by gas chromatography and selective detection is also used (Eisenbrand & Preussmann, 1970; Sen, 1970; Althorpe et al., 1970; Alliston et al., 1972).

3. Biological Data Relevant to the Evaluation of Carcinogenic Risk to Man

3.1 Carcinogenicity and related studies in animals

(a) Oral administration

Mouse: DBN was administered at a concentration of 50 ppm in the diet of ICR mice for 12 months. All of the 33 animals surviving for more than 12 months developed squamous cell carcinomas or papillomas of the forestomach. Furthermore, 15 animals developed liver tumours. Papillomas of the oesophagus and adenomas of the lung were observed in 4 and 8 mice, respectively. In the 28 control animals living more than 15 months, lung adenomas developed in 2 mice, and a lymphatic leukaemia was observed in 1 other mouse (Takayama & Imaizumi, 1969). In 2 groups of 50 male and 50 female C57BL/6 mice, continuous administration of 30 mg and 8 mg/kg bw/day in the drinking-water led to the development of squamous cell carcinomas and papillomas of the oeso-phagus in almost all experimental animals at both dosages, together with some tumours of the tongue, the soft palate and the pharynx. At the lower dosage carcinomas of the forestomach also occurred. While no liver tumours were seen in this study, 48% of the mice at the higher dose level and 21% at the lower dose level developed papillo-mas and multifocal squamous cell carcinomas of the urinary bladder. Male animals were more sensitive towards bladder carcinogenesis than were females (Bertram & Craig, 1970).

Rat: The carcinogenicity of DBN in the rat was first demonstra-ted by Druckrey et al. (1962, 1967). Four dose levels were investigated in BD rats. At a dose of 75 mg/kg bw/day in the diet liver carcinomas were induced in all 4 surviving rats. At the lower dosages (37.5, 20 and 10 mg/kg bw/day) the incidence of liver cancer decreased, whereas malignant tumours in the oesophagus and pharynx were observed. Squamous cell carcinomas of the urinary bladder occurred at the 2 middle dose levels in 5/16 and 7/10 animals. Similar results were reported in Wistar rats after the administration of 0.01 and 0.05% DBN in the drinking-water. Bladder tumours occurred at both dose levels, while liver cancers were seen only at the higher dose level (Okajima et al., 1970). Simultaneous administration of DL-tryptophane did not influence carcinogenesis in the bladder and oesophagus of male Wistar rats but completely inhibited the development of liver tumours (Okajima et al., 1971). In 30 female Wistar rats given 20 mg/kg bw/day DBN in the drinking-water, 16 papillomas and 2 carcinomas of the urinary bladder were found (Kunze & Shauer, 1971: Kunze et al., 1969, 1971).

Hamster: Groups of 100 Syrian golden and 66 Chinese hamsters given doses of 300 mg/kg bw/week DBN by stomach tube for life showed a low incidence of papillomas and carcinomas of the bladder. Syrian hamsters had a high incidence (41%) of respiratory tract tumours. On the other hand, the Chinese hamsters developed papillomas (59%) and carcinomas (32%) in the forestomach, but no tumours in the trachea and lungs (Althoff et al., 1971; Mohr et al., 1970). In Syrian golden hamsters, single doses of 400, 800 and 1600 mg/kg bw produced tumours of the respiratory tract in 3/10, 5/10 and 7/10 animals, respectively, the first tumour appearing after 31 weeks (Althoff et al., 1973).

Guinea pig: At a dosage of 40 mg/kg bw/day DBN, 5 times a week for life, all 15 animals surviving more than 550 days developed hepatocellular carcinomas; cholangiomas also appeared in a few animals. Three of the animals had papillomas and 4 had squamous cell carcinomas of the urinary bladder (Ivankovic & Bücheler, 1968).

(b) Subcutaneous and/or intramuscular administration

Mouse: Bi-weekly s.c. treatment of IF x C57 mice (30 males and
30 females) with 10 µl pure DBN per mouse for up to 40 weeks resulted in
severe haematuria and, subsequently, in the development of urinary
bladder carcinomas in 7/17 females and 1 papilloma and 6 carcinomas
in 7/12 males. Histologically the tumours were papillomas, transitional cell and anaplastic carcinomas, together with a few squamous
cell carcinomas (Wood et al., 1970).

Newborn mouse: Liver tumours developed in 36/37 male and 20/24
female IF x C57 mice given 1 µl pure DBN s.c. on days 1, 8 and 15 of life
A few lung adenomas and carcinomas were observed but no urinary
bladder tumours. No liver or lung tumours were observed in the controls (Wood et al., 1970).

Rat: Weekly s.c. injections of 400 and 200 mg/kg bw DBN to BD
rats for life resulted in the development of carcinomas of the urinary
bladder in 18/20 rats as well as carcinomas of the oesophagus in 3/20
and of the liver in 2/20 rats. S.c. administration thus leads to an
increased incidence of urinary bladder cancer as compared with oral
administration (Druckrey et al., 1967).

Hamster: A group of 25 Syrian golden hamsters treated s.c. with
300 mg/kg bw/week DBN developed a high incidence (84%) of tracheal
papillomas and a moderate incidence of urinary bladder papillomas
(16%) and carcinomas (24%). On the other hand, 16 Chinese hamsters,
similarly treated, developed forestomach papillomas (56%) and carcinomas (6%). Urinary bladder papillomas (6%) and carcinomas (12%) were
found, but no tumours developed in the respiratory tract (Mohr et al.,
1970; Althoff et al., 1971). Single s.c. doses of 150, 300, 600 and
1200 mg/kg bw given to 4 groups of Syrian golden hamsters (10 per
group) produced tumours of the respiratory tract in 3, 4, 5 and 7
animals, respectively, the first tumour appearing after 19 weeks
(Althoff et al., 1973).

(c) Intraperitoneal administration

Hamster: Single doses of 200, 400 and 800 mg/kg bw given to 3 groups of 5 male and 5 female Syrian golden hamsters produced tumours of the respiratory tract in 4, 6 and 7 animals respectively, the first tumour appearing after 17 weeks. Some papillary tumours of the fore-stomach were also seen (Althoff et al., 1973).

(d) Other experimental systems

Intravenous administration: Twice weekly injections of 6 µg DBN/animal for 25 to 30 weeks to 43 male and 46 female CBA/H-T$_6$H$_6$ mice resulted in acute leukaemia of the reticulum cell type in 13/39 males and 17/37 females within 92 to 218 days. No leukaemia was observed in 51 control animals (Emura, 1970).

3.2 Other relevant biological data

(a) Animals

DBN needs enzymatic activation in the mammalian organism to form the proximate and/or ultimate carcinogen (Magee & Barnes, 1967; Druckrey et al., 1967). Terminal hydroxylation of DBN to form N-butyl-N-(4-hydroxybutyl)nitrosamine as a urinary metabolite was first described by Druckrey et al. (1964). Recently, several metabolites of this compound have been identified in the urine of rats, including n-butyl-(3-carboxy-propyl)nitrosamine, n-butyl-(3-carboxy-2-hydroxy-propyl)nitrosamine and the glucuronide of butyl(4-hydroxy-butyl) nitrosamine (Okada & Suzuki, 1972; Blattmann & Preussmann, 1973). Oxidative N-debutylation to form butyraldehyde has been shown to occur in vitro with rat liver microsomal fractions (Blattmann & Preussmann, 1973). The alkylation of cellular macromolecules by 1-(^{14}C)-di-n-butylnitrosamine has been studied in vivo; its admini-stration led to the formation of 7-(^{14}C)-n-butylguanine and 7-(^{14}C)-methylguanine in liver RNA. However, when the rats were given 2-(^{14}C)-di-n-butyl-nitrosamine, only 7-(^{14}C)-n-butylguanine was found. These results suggest that β-oxidation of this aliphatic nitrosamine occurs in vivo with metabolic splitting between the α- and β-carbon atoms and the formation of methylating intermediates (Magee, 1968; Krüger, 1971, 1972).

(b) Carcinogenicity of metabolites

The carcinogenicity of the urinary metabolite n-butyl-(4-hydroxy-butyl)-nitrosamine (BBN):

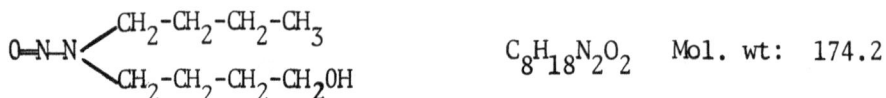

$$O=N-N \begin{cases} CH_2-CH_2-CH_2-CH_3 \\ CH_2-CH_2-CH_2-CH_2OH \end{cases} \qquad C_8H_{18}N_2O_2 \qquad \text{Mol. wt:} \quad 174.2$$

has been investigated. It has a boiling point of 115-116°C (at 0.01 mm) (Druckrey et al., 1964, 1967) and can be identified by thin-layer chromatography (Preussman et al., 1964). Approximately 2.5% BBN is soluble in water (Druckrey et al., 1967). BBN is stable at room temperature for more than 14 days in aqueous solution at neutral and alkaline pH in the absence of light (Druckrey et al., 1967).

Treatment of 50 male and 50 female C57BL/6 mice with 35 mg/kg bw/day BBN in the drinking-water for life induced anaplastic carcino-mas of the urinary bladder in all animals reaching autopsy with an average tumour induction time of 190 days for the males and 253 for the females. This sex difference in susceptibility can be abolished by castrating males or by treating females with testosterone (Bertram & Craig, 1972). The induction of carcinomas of the urinary bladder in mice has been described by Akagi et al. (1970).

In rats, daily oral doses of 40 and 20 mg/kg bw to 5 and 20 BD rats in the drinking-water selectively induced multiple haemorrhagic carcinomas of the urinary bladder in all animals (Druckrey et al., 1964). Identical results have been obtained by Ito et al. (1969) using Wistar rats and by Hashimoto et al. (1972) in ACI/N rats.

The incidence of tumours of the ureter, which were seen occasionally, in rats treated with BBN was greatly increased after

204

ligation of one ureter; this suggests that the concentration of the carcinogen is an important factor in the induction of tumours in the urinary system (Ito et al., 1971).

The influence of the length of treatment has been investigated in male Wistar rats to which 0.05% BBN was given in the drinking-water for 2, 4, 6, 8 and 12 weeks. At the time the animals were sacrificed (40 weeks) a clear dose-response effect (see table below) was observed in the induction of bladder papillomas and carcinomas (Ito et al., 1972).

Duration of treatment	% of animals with changes in the urinary bladder		
(weeks)	Hyperplasia	Papilloma	Carcinoma
2	66	33	0
4	100	72	18
6	94	88	64
8	100	90	90
12	100	100	100

Significant differences in the induction by BBN of urinary bladder tumours were seen in different strains of rats. The tumour incidence varied from 0% in Lewis rats to 100% in ACI/NC rats, with intermediate values in Sprague-Dawley, BDIX/N and Wistar rats (Ito et al., 1972). Co-administration of another carcinogenic nitrosamine, N-nitrosopiperidine, significantly reduced the yield of urinary bladder tumours (Ito et al., 1972).

The carcinogenicity of another metabolite of DMN and BBN, n-butyl-(3-carboxypropyl)nitrosamine (BCPN), has been demonstrated. It has the following formula:

$$O=N-N \Big\langle \begin{array}{l} CH_2-CH_2-CH_2-CH_3 \\ CH_2-CH_2-CH_2-\overset{\underset{\displaystyle O}{\|}}{C}-OH \end{array} \qquad C_8H_{16}N_2O_3 \quad \text{Mol. wt: } 188.1$$

A total of 8 ACI/N rats were treated with 0.06% BCPN in the drinking-water up to 20 weeks. Three animals killed at 14 weeks showed papillomas of the urinary bladder, and the remaining 5 animals which were killed at 28 weeks all showed transitional cell carcinomas (Hashimoto et al., 1972).

3.3 Observations in man

No data are available to the Working Group.

4. Comments on Data Reported and Evaluation[1]

4.1 Animal data

N-Nitroso-di-n-butylamine (DBNA) is carcinogenic in all animal species tested: mouse, rat, hamster and guinea pig. It is carcinogenic by oral and other routes, and it is particularly effective as a bladder carcinogen after subcutaneous injection. It is carcinogenic after exposure to a single dose.

4.2 Human data

No epidemiological data are available to the Working Group.

---

[1] See also the section "Extrapolation from animals to man" in the introduction to this volume.

5. References

Agaki, G., Agaki, A. & Kimura, M. (1970) Tumour of the urinary bladder induced by N-butyl-N-butanol-(4)-nitrosamine (BBN) in mice and rats. In: Proceedings of The Japanese Cancer Association 29th Annual Meeting, Tokyo, The Japanese Cancer Association, p. 65

Alliston, T.G., Cox, G.B. & Kirk, R.S. (1972) The determination of steam volatile N-nitrosamines in foodstuffs by the formation of electron capturing derivatives from electrochemically derived amines. Analyst, 97, 915

Althoff, J., Krüger, F.W., Mohr, W. & Schmähl, D. (1971) Dibutylnitrosamine carcinogenesis in Syrian golden and Chinese hamsters. Proc. Soc. exp. Biol. (N.Y.), 136, 168

Althoff, J., Pour, P., Cardesa, A. & Mohr, U. (1973) Comparative studies of neoplastic response to a single dose of nitroso compounds. II. The effect of N-dibutylnitrosamine in the Syrian golden hamster. Z. Krebsforsch., 79, 85

Althorpe, J., Goddard, D.A., Sissons, D.J. & Telling, G.M. (1970) The gas chromatographic determination of nitrosamines at the picogram level by conversion to their corresponding nitramines. J. Chromat., 53, 371

Bertram, J.S. & Craig, A.W. (1970) Induction of bladder tumours in mice with dibutylnitrosamine. Brit. J. Cancer, 24, 352

Bertram, J.S. & Craig, A.W. (1972) Specific induction of bladder cancer in mice by butyl-(4-hydroxybutyl)-nitrosamine and the effects of hormonal modifications on the sex difference in response. Europ. J. Cancer, 26, 515

Blattmann, L. & Preussmann, R. (1973) Struktur von Metaboliten carcinogener Dialkylnitrosamine im Rattenurin. Z. Krebsforsch., 79, 3

Bogovski, P., Preussmann, R. & Walker, E.A., eds., (1972) N-Nitroso Compounds, Analysis and Formation, Lyon, IARC Scientific Publications, 3

Burgess, E.M. & Lavanish, J.M. (1964) Photochemical decomposition of N-nitrosamines. Tetrahedron Letters, No. 20, 1221

Derr, P.F. (1960) Reduction of nitrosamines. US Patent 2, 961, 467 (Chem. Abstr., 55, 9280 i)

Druckrey, H., Preussmann, R., Ivankovic, S. & Schmähl, D. (1967) Organotrope carcinogene Wirkungen bei 65 verschiedenen N-Nitroso-Verbindungen an BD-Ratten. Z. Krebsforsch., 69, 103

Druckrey, H., Preussmann, R., Ivankovic, S., Schmidt, C.H., Mennel, H.D. & Stahl, K.W. (1964) Selektive Erzeugung von Blasenkrebs an Ratten durch Dibutyl- und N-Butyl-N-butanol (4) -nitrosamin. Z. Krebsforsch., 66, 280

207

Druckrey, H., Preussmann, R., Schmähl, D. & Müller, M. (1962)   Erzeugung von Blasenkrebs an Ratten mit N,N-Dibutylnitrosamin. Naturwissenschaften, 49, 19

Eisenbrand, G. & Preussmann, R. (1970)   Eine neue Methode zur kolori-metrischen Bestimmung von Nitrosaminen nach Spaltung der N-Nitroso-gruppe mit Bromwasserstoff in Eisessig. Arzneimittel-Forsch., 20, 1513

Emura, M. (1970)   Some histologic and cytologic features of leukemias induced by dibutylnitrosamine in CBA/H-T_6T_6 mice. Jap. J. Genetics, 45, 71

Ender, F. & Čeh, L. (1971)   Conditions and chemical reaction mechanisms by which nitrosamines may be formed in biological products with reference to their possible occurrence in food products. Z. Lebensmittel-Untersuch., 145, 133

Fiddler, W., Doerr, R.C., Ertel, J.R. & Wasserman, A.E. (1971)   Gas-liquid chromatographic determination of N-nitrosodimethylamine in ham. J. Ass. off. analyt. Chem., 54, 1160

Fridman, A.L., Mukhametshin, F.M. & Novikov, S.S. (1971)   Advances in the chemistry of aliphatic N-nitrosamines. Russ. chem. Rev., 40, 34

Gough, T.A. & Webb, K.S. (1972)   The use of a molecular separator in the determination of trace constituents by combined gas chromatography and mass spectrometry. J. Chromat., 64, 201

Gough, T.A. & Webb, K.S. (1973)   A method for the detection of traces of nitrosamines using combined gas chromatography and mass spectro-metry. J. Chromat., 79, 57

Hashimoto, Y., Suzuki, E. & Okada, M. (1972)   Induction of urinary bladder tumors in ACI/N rats by butyl(3-carboxypropyl)nitros[o]amine, a major urinary metabolite of butyl(4-hydroxybutyl)nitrosoamine. Gann, 63, 637

Hawksworth, G.M. & Hill, M.J. (1971)   Bacteria and the N-nitrosation of secondary amines. Brit. J. Cancer, 25, 520

Heyns, K. & Röper, H. (1971)   Ein spezifisches analytisches Trenn- und Nachweisverfahren für Nitrosamine durch Kombination von Capillar-gaschromatographie und Massenspektrometrie. Z. Lebensmittel-Untersuch., 145, 69

Ito, N., Hiasa, Y., Tamai, A., Okajima, E. & Kitamura, H. (1969)   Histo-genesis of urinary bladder tumors induced by N-butyl-N-(4-hydroxy-butyl) nitrosamine in rats. Gann, 60, 401

Ito, N., Hiasa, Y., Toyoshima, K., Okajima, E., Kamamoto, Y., Makiura, S., Yokota, Y., Sugihara, S. & Matayoshi, K. (1972) Rat bladder tumors induced by N-butyl-N-(4-hydroxybutyl) nitrosamine. In: Nakahara, W., Takayama, S., Sugimura, T. & Odashima, S.,eds., Topics in Chemical Carcinogenesis, Tokyo, University of Tokyo Press, p. 175

Ito, N., Makiura, S., Yokota, Y., Kamamoto, Y., Hiasa, Y. & Sugihara, S. (1971) Effect of unilateral ureter ligation on development of tumors in the urinary system of rats treated with N-butyl-N-(4-hydroxybutyl)-nitrosamine. Gann, 62, 359

Ivankovic, S. & Bücheler, J. (1968) Leber- und Blasen-Carcinome beim Meerschweinchen nach Di-n-butylnitrosamin. Z. Krebsforsch., 71, 183

Joffe, B.V. (1958) Synthesis of unsymmetric dialkylhydrazines. Zh. Obscej Khim., 28, 1296

Krüger, F.W. (1971) Metabolismus von Nitrosaminen in vivo. I. Uber die β-Oxidation aliphatischer Di-n-alkylnitrosamine: Die Bildung von 7-Methylguanin neben 7-Propyl-bzw. 7-Butylguanin nach Applikation von Di-n-propyl- oder Di-n-butyl-nitrosamin. Z. Krebsforsch., 76, 145

Krüger, F.W. (1972) New aspects in metabolism of carcinogenic nitrosamines. In: Nakahara, W., Takayama, S., Sugimura, T. & Odashima, S., eds.,Topics in Chemical Carcinogenesis, Tokyo, University of Tokyo Press, p. 213

Kunze, E. & Schauer, A. (1971) Enzymhistochemische und autoradiographische Untersuchungen an Dibutylnitrosamin-induzierten Harnblasenpapillomen der Ratte. Z. Krebsforsch., 75, 146

Kunze, E., Schauer, A. & Calvoer, R. (1969) Zur Histochemie von Harnblasen-Papillomen der Ratte, induziert durch Dibutylnitrosamin. Naturwissenschaften, 56, 639

Kunze, E., Schauer, A. & Spielmann, J. (1971) Autoradiographische Untersuchungen über den RNS-Stoffwechsel während der Entwicklung von Dibutylnitrosamin-induzierten Harnblasentumoren der Ratte. Z. Krebsforsch., 76, 236

Lijinsky, W. & Epstein, S.S. (1970) Nitrosamines as environmental carcinogens. Nature (Lond.), 225, 21

Magee, P.N. (1968) Possible mechanisms of carcinogenesis by N-nitroso compounds and alkylating agents. Fd Cosmet. Toxicol., 6, 572

Magee, P.N. & Barnes, J.M. (1967) Carcinogenic nitroso compounds. Advanc. Cancer Res., 10, 163

Mohr, U., Althoff, J., Schmähl, D. & Krüger, F.W. (1970) The carcinogenic effect of dibutylnitrosamine in Syrian and Chinese hamsters. Z. Krebsforsch., 74, 112

Okada, M. & Suzuki, E. (1972) Metabolism of butyl(4-hydroxybutyl)-nitrosoamine in rats. Gann, 63, 391

Okajima, E., Hiramatsu, T., Motomiya, Y., Iriya, R., Ijuin, M. & Ito, N. (1970) Bladder carcinogenesis with di-n-butylnitrosamine in the rat. In: Proceedings of the Japanese Cancer Association 29th Annual Meeting, Tokyo, The Japanese Cancer Association, p. 73

Okajima, E., Hiramatsu, T., Motomiya, Y., Iriya, R., Ijuin, M. & Ito, N. (1971) Effect of DL-tryptophan on tumorigenesis in the urinary bladder and liver of rats treated with N-nitrosodibutylamine. Gann, 62, 163

Preussmann, R. & Eisenbrand, G. (1972) Problems and recent results in the analytical determination of N-nitroso compounds. In: Nakahara, W., Takayama, S., Sugimura, T., Odashima, S., eds., Topics in Chemical Carcinogenesis, Tokyo, University of Tokyo Press, p. 323

Preussmann, R., Neurath, G., Wulf-Lorentzen, G., Daiber, D. & Hengy, H. (1964) Anfärbemethoden und Dünnschicht-Chromatographie von organischen N-Nitrosoverbindungen. Z. analyt. Chem., 202, 187

Rhoades, J.W. & Johnson, D.E. (1970) Gas chromatography and selective detection of N-nitrosamines. J. chromat. Sci., 8, 616

Sander, J. (1967) Kann Nitrit in der menschlichen Nahrung Ursache einer Krebsentstehung durch Nitrosaminbildung sein? Arch. Hyg. (Berlin), 151, 22

Sen, N.P. (1970) Gas-liquid chromatographic determination of dimethylnitrosamine as dimethylnitramine at picogram levels. J. Chromat., 51, 301

Takayama, S. & Imaizumi, T. (1969) Carcinogenic action of N-nitrosodibutylamine in mice. Gann, 60, 353

Telling, G.M., Bryce, T.A. & Althorpe, J. (1971) Use of vacuum distillation and gas chromatography-mass spectrometry for determination of low levels of volatile nitrosamines in meat products. J. agric. fd Chem., 19, 937

UICC (1970) The quantification of environmental carcinogens (UICC Technical Report Series, Vol. 4)

Wood, M., Flaks, A. & Clayson, D.B. (1970) The carcinogenic activity of dibutylnitrosamine in IF x C 57 mice. Europ. J. Cancer, 6, 433

Zsolnai, T. (1961) Versuche zur Entdeckung neuer Fungistatika. III. 8-Hydroxy-chinolin-Derivate, Nitroso-Verbindungen und Oxyme. Biochem. Pharmacol., 7, 195

N-NITROSO-N-METHYLURETHANE*

1. Chemical and Physical Data

1.1 Synonyms and trade names

Chem. Abstr. No.: 615-53-2

Ethyl ester of methylnitroso-carbamic acid; N-methyl-N-nitroso-ethylcarbamate; N-methyl-N-nitrosourethan; MNU; MNUN; NMUT

1.2 Chemical formula and molecular weight

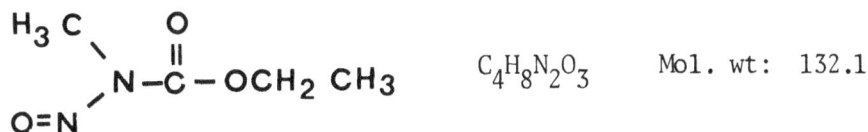

$$\underset{O=N}{\overset{\displaystyle H_3C}{\diagdown}} \!\! N \!\! - \!\! \overset{\displaystyle \overset{O}{\|}}{C} \!\! - \!\! OCH_2\,CH_3 \qquad\qquad C_4H_8N_2O_3 \qquad Mol.\ wt:\ 132.1$$

1.3 Chemical and physical properties of the pure substance

(a) Description: Yellow to pink oil of low viscosity

(b) Boiling-point: 62-64°C at 12 mm; 70°C at 27 mm; cannot be distilled at atmospheric pressure

(c) Density: d_4^{20} 1.133

(d) Refractive index: n_D^{20} 1.43632

(e) Absorption spectroscopy: λ_{max} 237 nm; log ε 3.77 (in water)

(f) Solubility: Only slightly soluble in water; soluble in many common organic solvents

---

* Considered by the Working Group in Lyon, June 1973.

(g) Stability: Unstable. It is sensitive to light, and some photodecomposition products have been described (Schoental, 1963b). Spontaneous decomposition may occur, especially on heating. Stability in aqueous solutions is pH-dependent (Druckrey et al., 1967). At alkaline pH, it decomposes to diazomethane (Tempe et al., 1964), which can also act as an alkylating agent. At $20^{o}C$, the following half-lives in hours at various pH values have been reported: 120 at pH 6.0; 80 at pH 7.0; 17.5 at pH 8.0; and 0.9 at pH 9.0. However, at $37^{o}C$ the following half-lives have been reported:

pH 6	pH 7	
58 (pH 5)	24	(McCalla et al., 1968)
50	10	(Schaper, 1970)
76	18.3 (pH 7.27)	(Lawley, 1968)

The differences may be explained by different experimental conditions, especially the type of buffer used.

(h) Volatility: Extremely volatile

(i) Chemical reactivity: N-Nitroso-N-methylurethane (NMUT) is highly reactive. It reacts easily with various nucleophiles, especially with thio groups (Schoental, 1961; Schoental & Rive, 1963, 1965; Schaper, 1970).

1.4 Technical products and impurities

No data on technical products and impurities are available to the Working Group.

2. Production, Use, Occurrence and Analysis

(a) Production and use[1]

NMUT can be made by the nitrosation of N-methylurethane. No evidence was found that it is produced commercially except as a research chemical or that there is any human exposure except for that of researchers engaged in the study of its biological properties.

NMUT was among the first N-nitroso compounds to be used as a raw material for the laboratory preparation of diazomethane solutions, but it has long since been displaced by other chemicals in this application; N-nitrosomethyl-p-toluenesulphonamide is the preferred reagent today (Druckrey & Preussmann, 1962).

(b) Occurrence

NMUT can be formed from N-methylurethane and nitrosating agents (such as nitrite) under conditions prevalent in the human stomach (Sander et al., 1971). The kinetics of the nitrosation reaction have been determined (Mirvish, 1971). No data are available on the occurrence of N-methylurethane in the human environment, but nitrite and other nitrosating agents are found almost ubiquitously in the environment.

(c) Analysis

No specific analytical method is available at present. As with other nitrosamides, it can be determined by colorimetry after photochemical splitting (Daiber & Preussmann, 1964) or acid splitting (Schaper, 1970; Preussmann & Schaper-Druckrey, 1972).

---

[1] Data from Chemical Information Services, Stanford Research Institute, USA.

3. Biological Data Relevant to the Evaluation of Carcinogenic Risk to Man

3.1 Carcinogenicity and related studies in animals

(a) Oral administration

Mouse: In a small number of animals (Schoental, 1963a) and in another study of unspecified size (Schoental, 1968), oral administration of NMUT was reported to have induced squamous cell papillomas and carcinomas of the oesophagus and forestomach, and adenocarcinomas of the glandular stomach. Control data were not reported.

Rat: Twenty BD strain rats were given NMUT in a daily dose of 4 mg/kg bw in the drinking-water for 330 days. Squamous cell carcinomas of the forestomach developed in 10 animals (Druckrey et al., 1961, 1967). The histogenesis of this type of tumour from day 1 to day 468 in rats given 0.01% NMUT continuously in the drinking-water has been described (Toledo, 1965). In this experiment, 1 papilloma and 1 squamous cell carcinoma of the oesophagus were also seen. The carcinogenicity of this compound for the upper gastro-intestinal tract has been confirmed in other studies (Schoental & Magee, 1962; Schoental, 1963a), and occasional adenomas and adenocarcinomas of the glandular stomach have also been reported (Schoental, 1968).

Hamster: Weekly intragastric doses of 0.8 mg NMUT per animal to 16 Syrian hamsters for 2 months and then twice-weekly doses for 4 to 5 months induced 13 epidermoid carcinomas of the oesophagus and epidermoid carcinomas of the forestomach in all treated animals (Herrold, 1966).

Guinea-pig: Continuous administration of 2.5 mg/kg bw NMUT in the drinking-water, 5 days per week for life gave rise to tumours in 27/38 animals. Adenocarcinomas were found in the stomach (pyloric and cardiac regions) and pancreas. A few tumours were also found in the lungs and larynx. Many of the tumours induced had metastases (Druckrey et al., 1967, 1968; Bücheler & Thomas, 1971).

(b) Subcutaneous and/or intramuscular administration

Mouse: Suckling dd/I strain mice 7 days of age were given s.c.
3.3 mg/kg bw NMUT per month for life. No tumours were found at the
site of injection. Pulmonary adenomas were found in all animals
surviving more than 80 days, while adenocarcinomas or anaplastic
carcinomas developed in 26/32 mice surviving that long. In 15/69
untreated controls adenomas, but no carcinomas, were observed.
Furthermore, a few lymphomas and ovarian tumours were induced. No
tumours were induced under similar experimental conditions with
C57BL/6/MS and CBA-T_6T_6/H strain mice within 13 months (Matsuyama et
al., 1969). Early biochemical changes in the lungs and ultrastructural characteristics of the tumours have been described
(Caulet & Pluot, 1970; Pluot et al., 1972).

Hamster: Weekly s.c. injections of 0.2 mg NMUT to 12 Syrian
hamsters for 5 to 6 months induced a widespread spectrum of pathological changes in the lungs. Three pulmonary adenomas and 1
anaplastic adenocarcinoma were seen in the lungs of animals surviving
7 to 20 months (Herrold, 1967).

Guinea-pig: Weekly s.c. injections of 3 mg/kg bw NMUT given to
5 animals for 1 year led to 2 fibrosarcomas at the injection site.
No distant tumours were seen (Druckrey et al., 1968).

(c) Intraperitoneal administration

Newborn mouse: A single dose of 0.08 mM (10.5 mg)/kg bw NMUT
was given to a group of 36 male and female newborn Swiss mice. Lung
adenomas developed within 365 days in all treated animals (3.3
tumours per mouse), compared with 0.06 tumours per mouse in 50 controls. One lymphoma was also seen, together with 1 carcinoma of the
lung. Lung tumourigenesis was more effective in newborn animals than
in adult mice in which a single dose of 0.19 mM (25 mg)/kg bw produced only 0.86 lung adenomas/mouse (Frei, 1970).

Rat: A single dose of 5 mg NMUT/rat (50 mg/kg bw), or two doses
of 5 mg NMUT/rat at an interval of 5 months, produced adenomas and

adenocarcinomas in the ileum, caecum or rectum in 7 rats dying 5 to 23 months after the first injection. Pulmonary adenomatosis occurred in 9 rats, with 1 adenocarcinoma and 1 carcinoma in 2 of these rats; 1 tumour of the spleen and uterus and a mesothelioma and a carcinoma of the seminal vesicle were also observed in animals dying up to 2 years. No control animals were used (Schoental & Bensted, 1968).

(d)   Other experimental systems

Intravenous administration: Repeated i.v. injections to 20 rats of 1 mg/kg bw NMUT every 9 days (total dose, 40 mg/kg bw) produced malignant lung tumours, diagnosed as alveolar cell and squamous cell carcinomas, in 18 rats dying up to 400 days (Druckrey et al., 1962, 1967; Thomas & Schmähl, 1963).

NMUT injected i.v. in doses of 4 mg/kg bw 3 times per week for 3 months induced papillary or adenomatous lung tumours in 3/18 rabbits (Itano et al., 1972).

Pre-natal exposure: In rats a dose of 0.1 mg NMUT on the 21st day of pregnancy gave tumours in the offspring with an incidence of 56%. Tumours were localized in the nervous system (n.trigeminus), kidney, lungs, mammary and adrenal gland and thyroid. No early induction of tumours was seen in the offspring when 40 mg/kg bw NMUT was given intraperitoneally or 10 mg/kg bw i.v. to the parent females on day 21 of pregnancy (Alexandrov, 1972). In another experiment (Tanaka, 1973) NMUT was given i.v. at a dose of 40 mg/kg bw to WKA female rats at various periods of gestation. An increased incidence of tumours, including tumours of the nervous tissue, was observed in the offspring, the highest incidence occurring when NMUT was given between days 14 and 18 of gestation. An increased incidence of tumours was reported to occur also in the second generation of un-treated descendants of the NMUT-treated mothers.

3.2 Other relevant biological data

(a) Animals

NMUT is highly reactive chemically and can combine with nucleic acids and proteins in vitro without enzymatic activation (Schoental, 1961; Schoental & Rive, 1963). Methylation of nucleic acids in vivo and in vitro has been demonstrated; the main reaction product is 7-methylguanine (Schoental, 1967), and 3-methyladenine has been found to be another (Lawley, 1968). Methylation also occurs in organs that are not usually cancerized by NMUT (Schoental, 1969). Decomposition of NMUT is accelerated by cysteine in vitro, forming S- and carboxyl-methylated and S- and N-ethoxycarbonylated cysteines (Schoental & Rive, 1965).

3.3 Observations in man

No data are available to the Working Group.

4. Comments on Data Reported and Evaluation[1]

4.1 Animal data

N-Nitroso-N-methylurethane (NMUT) is carcinogenic in all species tested: mouse, rat, hamster, guinea-pig. It has a local as well as a systemic carcinogenic effect, producing tumours at different sites by oral and other routes of administration. It is carcinogenic in single-dose experiments and following pre-natal exposure.

4.2 Human data

No epidemiological data are available to the Working Group.

---

[1] See also the section "Extrapolation from animals to man" in the introduction to this volume.

5. References

Alexandrov, V.A. (1972) On the effect of N-nitrosomethylurethan on rat
   embryos. Vop. Onkol., 18, 59

Bücheler, J. & Thomas, C. (1971) Experimentelle erzeugte Drüsenmagen-
   tumoren bei Meerschweinchen und Ratte. Beitr. Path., 142, 194

Caulet, T. & Pluot, M. (1970) Etude histologique et histochimique des
   lésions induites au niveau du poumon de souris par un carcinogène:
   le N-nitroso-N-methylurethane. Intérêt de l'exploration des
   hydrolases. Z. Krebsforsch., 74, 227

Daiber, D. & Preussmann, R. (1964) Quantitative colorimetrische
   Bestimmung organischer N-Nitroso-Virbindungen durch photochemische
   Spaltung der Nitrosaminbindung. Z. analyt. Chem., 206, 344

Druckrey, H., Ivankovic, S., Bücheler, J., Preussmann, R. & Thomas, C.
   (1968) Erzeugung von Magen- und Pankreas-Krebs beim Meerschweinchen durch Methylnitroso-harnstoff und -urethan. Z.Krebsforsch.,
   72, 167

Druckrey, H. & Preussmann, R. (1962) N-nitroso-N-methylurethane: a
   potent carcinogen. Nature (Lond.), 195, 1111

Druckrey, H., Preussmann, R., Afkham, J. & Blum, G. (1962) Erzeugung von
   Lungenkrebs durch Methylnitrosourethan bei intravenöser Gabe an Ratten.
   Naturwissenschaften, 49, 451

Druckrey, H., Preussmann, R., Ivankovic, S. & Schmähl, D. (1967) Organotrope carcinogene Wirkungen bei 65 verschiedenen N-Nitroso-Verbindungen an BD-Ratten. Z. Krebsforsch., 69, 103

Druckrey, H., Preussmann, R., Schmähl, D. & Müller, M. (1961) Erzeugung
   von Magenkrebs durch Nitrosamide an Ratten. Naturwissenschaften, 48,
   165

Frei, J.V. (1970) Toxicity, tissue changes, and tumor induction in inbred
   Swiss mice by methylnitrosamine and -amide compounds. Cancer Res.,
   30, 11

Herrold, K. McD. (1966) Epidermoid carcinomas of esophagus and forestomach induced in Syrian hamsters by N-nitroso-N-methylurethan.
   J. nat. Cancer Inst., 37, 389

Herrold, K. McD. (1967) Fibrosing alveolitis and atypical proliferative
   lesions of the lung. An experimental study in Syrian hamsters.
   Amer. J. Path., 50, 639

Itano, T., Takimoto, R., Konishi, T. & Kagawa, T. (1972) Occurrence of lung cancer of rabbit induced by administration of N-methyl-N-nitrosourethane. Igaku No Ayumi, 81, 826

Lawley, P.D. (1968) Methylation of DNA by N-methyl-N-nitrosourethane and N-methyl-N-nitroso-N'-nitroguanidine. Nature (Lond.), 218, 580

Matsuyama, M., Suzuki, H. & Nakamura, T. (1969) Carcinogenesis in dd/I mice injected during suckling period with urethane, nitrogen mustard N-oxide and nitrosourethane. Brit. J. Cancer, 23, 167

McCalla, D.R., Reuvers, A. & Kitai, R. (1968) Inactivation of biologically active N-methyl-N-nitroso compounds in aqueous solution: effect of various conditions of pH and illumination. Canad. J. Biochem., 46, 807

Mirvish, S.S. (1971) Kinetics of nitrosamide formation from alkylureas, N-alkylurethane and alkylguanidines: possible implications for the etiology of human gastric cancer. J. nat. Cancer Inst., 46, 1183

Pluot, M., Hopfner, C., Adnet, J.J. & Chaulet, T. (1972) Aspect ultrastructural des lésions induites au niveau du poumon de souris par le N-nitroso-N-méthylurethane. Essai d'interprétation des inclusions cristallines cellulaires. Z. Krebsforsch., 77, 279

Preussmann, R. & Schaper-Druckrey, F. (1972) Investigation of a colorimetric procedure for determination of nitrosamides and comparison with other methods. In: Bogovski, P., Preussmann, R. & Walker, E.A., eds., N-Nitroso Compounds, Analysis and Formation, Lyon, IARC Scientific Publications, 3, p. 81

Sander, J., Bürkle, G., Floche, L. & Aeikens, B. (1971) Untersuchungen in vitro über die Möglichkeit einer Bildung cancerogener Nitrosamide im Magen. Arzneimittel-Forsch., 21, 411

Schaper, F. (1970) Zur Analytik von N-Alkyl-N-Nitrosamiden (Thesis, University of Freiburg/Br.)

Schoental, R. (1961) Interaction of the carcinogenic N-methyl-nitrosourethane with sulphydryl groups. Nature (Lond.), 192, 670

Schoental, R. (1963a) Experimental induction of squamous carcinoma of the lung, oesophagus and stomach, the mode of their induction. Acta Un. int. Cancr., 19, 680

Schoental, R. (1963b) Photodecomposition of N-alkyl-N-nitrosourethanes. Nature (Lond.), 198, 1089

Schoental, R. (1967) Methylation of nucleic acids by $N(^{14}C)$-methyl-N-nitrosourethane in vitro and in vivo. Biochem. J., 102, 5C

Schoental, R. (1968) Experimental induction of gastro-intestinal tumors in rodents by N-alkyl-N-nitrosourethans and certain related compounds. Gann Monogr., 3, 61

Schoental, R. (1969) Lack of correlation between the presence of 7-methylguanine in deoxyribonucleic acid and ribonucleic acid of organs and the localization of tumors after a single carcinogenic dose of N-methyl-N-nitrosourethane. Biochem. J., 114, 55P

Schoental, R. & Bensted, J.P.M. (1968) Tumors of the intestines induced in rats by intraperitoneal injections of N-methyl- and N-ethyl-N-nitrosourethanes. Brit. J. Cancer, 22, 316

Schoental, R. & Magee, P.N. (1962) Induction of squamous carcinoma of the lung and of the stomach and oesophagus by diazomethane and N-methyl-N-nitrosourethane, respectively. Brit. J. Cancer, 16, 92

Schoental, R. & Rive, D.J. (1963) The interaction of the carcinogen N-methyl-N-nitrosourethane with cysteine in vitro. Biochem. J., 87, 22P

Schoental, R. & Rive, D.J. (1965) Interaction of N-alkyl-N-nitrosourethanes with thiols. Biochem. J., 97, 466

Tanaka, T. (1973) Transplacental induction of tumours and malformations in rats treated with some chemical carcinogens. In: Tomatis, L. & Mohr, U., eds.,Transplacental Carcinogenesis, Lyon, IARC Scientific Publications, 4, p. 100

Tempé, J., Heslot, H. & Morel, G. (1964) Etude de l'hydrolyse alcaline de quelques dérivés N-nitrosés, précurseurs de diazoalcanes. C.R. Acad. Sci. (Paris), 258, 5470

Thomas, C. & Schmähl, D. (1963) Zur Morphologie der durch intravenöse Injektion von Nitrosomethylurethan erzeugten Lungentumoren bei der Ratte. Z. Krebsforsch., 65, 294

Toledo, J.D. (1965) Die Cytogenese des Vormagencarcinoms der Ratte durch N-methyl-N-Nitrosourethan. Beitr. Path. Anat., 131, 63

STREPTOZOTOCIN*

1. Chemical and Physical Data

1.1 Synonyms and trade names

Chem. Abstr. No.: 18883-66-4

N-D-Glucosyl-(2)-N'-nitrosomethylurea; 2-deoxy-2-(3-methyl-3-nitro-soureido)-D-glucopyranose; STR

1.2 Chemical formula and molecular weight

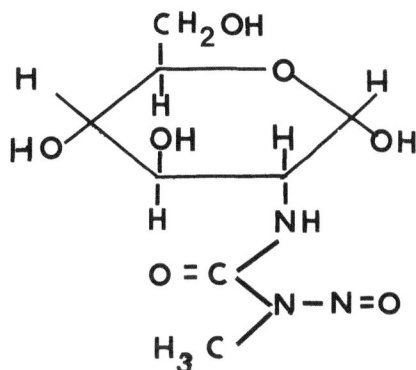

$C_8H_{15}N_3O_7$ Mol. wt: 265

1.3 Chemical and physical properties of the pure substance

(a) Description: White to pale yellow crystals

(b) Melting-point: $115^{o}C$ (decomposition)

---

* Considered by the Working Group in Lyon, June 1973.

(c) Absorption spectroscopy: λ_{max}228 nm;   log ε 6360

other λ_{max}380, 394 and 412 nm

(d) Optical rotation: α_D^{25} = + 39° in aqueous solution

(e) Solubility: Readily soluble in water and lower alcohols; only slightly soluble in non-polar organic solvents

(f) Stability: The pure compound decomposes rapidly at 70°C, but it is stable at 4°C for more than six months. As with other nitrosamides, the pure compound is sensitive to humidity and light. In aqueous solution, maximal stability is at pH 4; at lower and higher pH's, rapid decomposition occurs. At alkaline pH, it decomposes to form diazomethane.

(g) Chemical reactivity: It can be acetylated to form tetraacetyl-streptozotocin (m.p., 111-114°C).

1.4 Technical products and impurities

Microbiologically-formed streptozotocin (STR) is often impure and various samples have different impurities.

2. Production, Use, Occurrence and Analysis

(a) Production and use[1]

STR was first isolated as an antibiotic (Vavra et al., 1960; Lewis & Barbiers, 1960). It can be obtained by isolation from Streptomyces achromogenes fermentation broth. It also has been synthesized by three different procedures: (i) from tetra-O-acetyl glucosamine hydrochloride (Herr et al., 1967); (ii) from D-glucosamine + N-nitrosomethyl carbamyl-azide (Hardegger et al., 1969); and (iii) from D-glucosamine N-methylurea (Hessler & Jahnke, 1970). It is produced commercially in the United States by one manufacturer of research biochemicals. This company offers

---

[1] Date from Chemical Information Services, Stanford Research Institute, USA.

222

very small quantities (250 mg-5 g) for sale.

STR was of interest as a potential antimicrobial agent, but it has never found commercial usage in this application. It is presently of research interest for use in studies of diabetes since it has a specific toxic action on β-cells of the pancreas, inducing hyperglycemia (Rakieten et al., 1963); it also has a cytotoxic effect against several experimental tumours in animals (White, 1963; Schein et al., 1967). It can be used clinically in the treatment of tumours of the β-cells of the pancreas (Rudas, 1972).

(b) Occurrence

STR is produced by the soil microorganism, Streptomyces achromogenes (Var. 128), first isolated from soil samples collected near Blue Rapids, Kansas, USA.

(c) Analysis

A colorimetric determination by denitrosation with dilute aqueous acids has been described (Forist, 1964) and improved (Schaper, 1970). The lower detection limit in blood is 3 µg/ml, and in tissues 5 µg/ml. Polarographic determination is also possible (Garrett, 1960).

3. Biological Data Relevant to the Evaluation of Carcinogenic Risk to Man

3.1 Carcinogenicity and related studies in animals

(a) Intravenous administration

Rat: Following a single i.v. injection of 50 mg/kg bw STR to adult male Holtzman rats, 10/19 rats survived for longer than 8 months, 9 rats developed adenomas of the renal cortex and 1 rat a tumour of the pancreas (Arison & Feudale, 1967). Of 37 animals treated in the same way, 6 males and 6 females survived over 8 months, and 5 males and 1 female had kidney tumours. Experiments designed to show that STR itself was carcinogenic and that an accompanying substance, also derived from Streptomyces achromogenes, zedalan, was not responsible for the observed tumourigenicity, gave similar results;

an 85:15 mixture of STR and zedalan was used. Tumours appeared between 304 and 485 days after the start of the experiment. Tumours were not seen in rats treated with zedalan alone nor in the solvent controls (Rakieten et al., 1968). Co-administration of 2 doses of nicotinamide (350 mg/kg bw) i.p. after a single dose of 50 mg/kg bw STR to 28 rats induced pancreatic islet cell tumours (nesidioblasto- mas) in 18/26 rats surviving longer than 8 months. STR alone produced a single pancreatic tumour in 1/21 animals surviving for 8 months or longer. No kidney tumours were observed. The nicotinamide group and solvent controls did not develop tumours in 550 days, by which time all experimental animals were dead (Rakieten et al., 1971).

Hamster: Of 35 animals injected with STR, 13 developed tumours of the liver. The tumours affected mainly the parenchyma (hepatomas) and the biliary epithelium, but 3 sarcomas were also seen in the liver, possibly of extrahepatic origin. The dose and period of ob- servation are not recorded (Sibay & Hayes, 1969).

3.2 Other relevant biological data

(a) Animals

In a review of the pharmacology of STR, White (1963) stated that tissue distribution had been studied in mice, rats, cats, monkeys and dogs. In all these species, STR given parenterally is markedly con- centrated in the liver and kidney; for example, in dogs, STR is retained in the liver for many hours after it can no longer be detected in the blood. In mice, STR is well absorbed from the gastrointestinal tract, but poorly so by monkeys and not all by dogs.

3.3 Observations in man

STR is used clinically for the treatment of pancreatic islet cell tumours Data for acute and subchronic toxic effects have been summarized (Rudas, 1972).

4.   Comments on Data Reported and Evaluation[1]

4.1   Animal data

Streptozotocin (STR) is carcinogenic in rats and hamsters following intravenous injection, the only route tested. It is active in single dose experiments.

4.2   Human data

No epidemiological data are available to the Working Group.

---

[1] See also the section "Extrapolation from animals to man" in the introduction to this volume.

5. References

Arison, R.N. & Feudale, E.L. (1967) Induction of renal tumour by streptozotocin in rats. Nature (Lond.), 214, 1254

Forist, A.A. (1964) Spectrophotometric determination of streptozotocin. Analyt. Chem., 36, 1338

Garrett, E.R. (1960) Prediction of stability in pharmaceutical preparations. VII. The solution degradation of the antibiotic streptozotocin. J. amer. Pharm. Ass., 49, 767

Hardegger, E., Meier, A. & Stoos, A. (1969) Eine präparative Synthese von Streptozotocin. Helv. Chim. Acta, 52, 2555

Herr, R.R., Jahnke, H.K. & Argoudelis, A.D. (1967) The structure of streptozotocin. J. amer. Chem. Soc., 89, 4808

Hessler, E.J. & Jahnke, H.K. (1970) Improved synthesis of streptozotocin. J. org. Chem., 35, 245

Lewis, C. & Barbiers, A.R. (1960) Streptozotocin, a new antibiotic. In vitro and in vivo evaluation. In: Welch, H. & Marti-Ibañez, F., eds., Antibiotics Annual, New York, Medical Encyclopedia, 7, p. 247

Rakieten, N., Gordon, B.S., Beaty, A., Cooney, D.A., Davis, R.D. & Schein, P.S. (1971) Pancreatic islet cell tumors produced by the combined action of streptozotocin and nicotinamide. Proc. Soc. exp. Biol. (N.Y.), 137, 280

Rakieten, N., Gordon, B.S., Cooney, D.A., Davis, R.D. & Schein, P.S. (1968) Renal tumorigenic action of streptozotocin (NSC-85998) in rats. Cancer Chemother. Rep., 52, 563

Rakieten, N., Rakieten, M.L. & Madkarni, M.V. (1963) Studies on the diabetogenic action of streptozotocin (NSC-37917). Cancer Chemother. Rep., 29, 91

Rudas, B. (1972) Streptozotocin. Arzneimittel-Forsch., 22, 830

Schaper, F. (1970) Zur Analytik von N-Alkyl-N-nitrosamiden. (Thesis, University of Freiburg/Br).

Schein, P.S., Cooney, D.A. & Vernon, M.L. (1967) The use of nicotinamide to modify the toxicity of streptozotocin diabetes without loss of anti-tumor activity. Cancer Res., 27, 2324

Sibay, T.M. & Hayes, J.A. (1969) Potential carcinogenic effect of streptozotocin. Lancet, ii, 912

Vavra, J.J., De Boer, C., Dietz, A., Hanka, L.J. & Sokolski, W.T. (1960) Streptozotocin, a new antibacterial antibiotic. In: Welch, H. & Marti-Ibañez, F., eds., Antibiotics Annual, New York, Medical Encyclopedia, 7, p. 230

White, F.R. (1963) Streptozotocin. Cancer Chemother. Rep., 30, 49

MISCELLANEOUS
ALKYLATING AGENTS

BIS(CHLOROMETHYL)ETHER[*]

1. Chemical and Physical Data

1.1 Synonyms and trade names

Chem. Abstr. No.: 432-88-1

BCME; Chloromethyl ether[1]; chloro(chloromethoxy)methane; dichloro-methyl ether; symmetrical-dichloro-dimethyl ether; symmetrical-dichloromethyl ether; dimethyl-1,1'-dichloroether

1.2 Chemical formula and molecular weight

$ClCH_2OCH_2Cl$ $C_2H_4Cl_2O$ Mol. wt: 115.0

1.3 Chemical and physical properties of the pure substance

(a) Description: Colourless liquid

(b) Boiling-point: $104^\circ C$ at 760 mm

(c) Density: d_4^{15} 1.328

(d) Refractive index: n_D^{20} 1.435

(e) Solubility: Miscible with ethanol, ether and many other organic solvents

(f) Volatility: Highly volatile

(g) Stability: Decomposes in the presence of water into hydrochloric acid and formaldehyde. About 70% of bis(chloromethyl)ether (BCME) is hydrolyzed in heavy water (D_2O) in two minutes and 80% after 18 hours, suggesting the attainment of an equilibrium between bis(chloromethyl)ether and its hydrolysis products (Van Duuren et al., 1969). It was recently reported that BCME at 10 and 100 ppm is stable in air with 70% relative humidity for at least 18 hours (Collier, 1972)

---

[*]Considered by the Working Group in Lyon, June 1973.

[1]This name is sometimes used in error for chloromethyl methyl ether.

1.4 Technical products and impurities

It has been reported that one grade of "commercial" BCME was >99% pure and contained a trace of carbonyl-containing impurity (Van Duuren et al., 1968).

2. Production, Use, Occurrence and Analysis

(a) Production and use[1]

BCME can be produced by saturating a solution of paraformaldehyde in cold sulphuric acid with hydrogen chloride.

Although BCME may be produced in the United States as a chemical intermediate, no evidence was found that it is produced on a commercial scale for sale to others. Several companies which sell laboratory chemicals offer it for sale, but apparently these companies merely serve as repackagers or distributors (the source of the chemical they repackage is not known). In 1967, one manufacturer of laboratory and experimental chemicals in the Federal Republic of Germany was reported to be offering BCME for sale. No information is available on the production of BCME in other countries, but it seems likely that it is produced at least as a chemical intermediate.

Alpha-halo ethers such as BCME were reported to have proved useful in the laboratory synthesis of other organic chemicals, and 30 or 40 patents have appeared since 1940 describing the preparation of various textile aids from chloromethyl ether with amides, amines, alcohols and other reagents (Summers, 1955). Such textile aids are not necessarily prepared from previously isolated halomethyl ethers.

In 1968, it was reported that at least 28 patents had appeared in Chemical Abstracts during the last decade concerning the synthesis and applications of chloromethyl methyl ether and bis(chloromethyl)ether. In spite of this apparent interest in BCME, sales in the US appear to have been limited to laboratory quantities (Van Duuren et al., 1968).

---

[1]Data from Chemical Information Services, Stanford Research Institute, USA.

The largest potential source of human exposure to BCME appears to be in the use of a chloromethylating reaction mixture made from methanol, formaldehyde and hydrochloric acid. BCME can be a major component of these reaction mixtures, which are used without isolation of the individual products for chloromethylating a variety of organic compounds. The chloromethylated products are primarily chemical intermediates which are often subsequently converted to aminated compounds by reaction with organic amines. Probably the largest single usage of chloromethylation is in the preparation of anion exchange resins. A modified polystyrene resin is chloromethylated and then treated with a tertiary amine or with a polyamine. These resins have been produced in the US, and their production in 1966 in the Federal Republic of Germany, in the German Democratic Republic, France, Italy, Japan, The Netherlands, United Kingdom and USSR has also been reported (Abrams & Benezra, 1967). However, no data are available on the quantities of resins produced.

Other chloromethylated compounds of commercial significance include chloromethyl diphenyl oxide, 1-chloromethylnaphthalene, di(chloromethyl) diphenyl oxide, di(chloromethyl)toluene and dodecylbenzyl chloride. No data are available on the total production of these chemicals, since they are largely used captively.

(b) Occurrence

BCME does not occur as such in nature. Because it hydrolyzes in water, it would not be expected to remain as such for protracted periods in waste streams from plants where it is produced or used. However, in view of its reported stability in moist air, it might be present in exhaust gases from such plants (Collier, 1972).

BCME is listed together with 13 other chemicals in the US Federal Register (US Government, 1973) as being subject to an Emergency Temporary Standard on certain carcinogens under an order made by the Occupational Safety and Health Administration, Department of Labor on 26 April, 1973.

<u>(c)</u> <u>Analysis</u>

A high-resolution mass spectral procedure which can measure BCME at
the 0.1 ppb level in air containing other organic compounds has been developed (Collier, 1972). Gas chromatography can also be used. Several
methods for the detection of alkylating agents have been reported (Preussmann
et al., 1969; Sawicki & Sawicki, 1969).

3. <u>Biological Data Relevant to the Evaluation</u>
<u>of Carcinogenic Risk to Man</u>

3.1 <u>Carcinogenicity and related studies in animals</u>

<u>(a) Skin application</u>

<u>Mouse</u>: As summarized by Van Duuren et al. (1972), BCME is a
powerful cutaneous carcinogen. Complete carcinogenic activity was
revealed in a test in which 2 mg BCME dissolved in 0.1 ml benzene were
applied to the skin of female ICR/Ha Swiss mice 3 times per week for
a total of 325 days. Papillomas developed in 13/20 mice, 12 of which
progressed to squamous cell carcinomas by 325 days. The first papilloma was seen at 161 days, the first carcinoma at 231 days. In an
initiation-promotion experiment, application to the skin of a single
dose of 1 mg BCME dissolved in 0.1 ml benzene, followed by thrice-
weekly applications of an acetone solution of 0.025 mg in 0.1 ml mixed
phorbol esters from croton oil 14 days after initiation treatment,
yielded papillomas in 5/20 female mice, the first papilloma being
noted at 76 days. Two mice had squamous cell carcinomas, 1 with
metastases to the lung. The median survival time was 474 days. No
tumours were seen in a control group which received BCME as initiator
and no promoting treatment. In mice receiving the mixed phorbol esters
alone, papillomas developed in 2/20 mice, the first appearing at 322
days. In a test involving pre-treatment with 0.15 mg benzo(a)pyrene
in 0.1 ml benzene solution, followed 2 weeks later by thrice-weekly
applications of BCME (2 mg in 0.1 ml phorbol esters), as in the
complete carcinogenicity experiment, papillomas developed in 13/20
mice, 12 of which progressed to carcinomas. The latent period of the
first papilloma was 98 days, shorter than in the test without benzo(a)

234

pyrene pre-treatment. However, the median survival time of 315 days was almost identical (Van Duuren et al., 1968, 1969, 1972).

(b) Inhalation and/or intratracheal administration

Mouse: Nominal concentrations of 0.005 mg/l (5 mg/m^3) BCME were introduced into a chamber containing 50 strain A/HE mice for 6 hours per day on 5 days per week, for a total of 82 exposure days in 27 weeks. Twenty-six of the surviving 47 animals exhibited lung tumours, with an average of 2.9 tumours per mouse. In untreated controls held under the same conditions for 130 days in a chamber and maintained for a total of 28 weeks 20/49 animals had lung tumours (with a multiplicity of 0.8 tumours per animal). In a positive control group exposed to 0.5 mg/l (500 mg/m^3) urethane for 130 days 46/49 animals had lung tumours (with a multiplicity of 54 tumours per animal) (Leong et al., 1974).

Rat: Laskin et al (1971) exposed Sprague-Dawley rats to several dose levels of BCME in inhalation chambers for 6 hours per day on 5 days per week. A preliminary report details findings at a low exposure level (0.1 ppm, or 0.5 mg/m^3) for a total of 101 exposures. After 659 days, all 30 animals at risk were dead. Of 19 animals in which pathological examination was completed at the time of reporting, 5 died between 332 and 463 days after the beginning of the test with squamous cell carcinomas of the lung. In 5 others, dying between 346 and 488 days after the beginning of the test, tumours of the nasal cavity (diagnosed as aesthesioneuro-epitheliomas invading the sinuses, cranial vault and brain) were noted. In the same report, mention is made of the occurrence of multiple tumours in the lung and nasal cavity in rats exposed for shorter periods, but details are not reported.

(c) Subcutaneous and/or intramuscular administration

Newborn mice: Male and female ICR Swiss random bred mice were given a single s.c. injection of a pre-determined maximum tolerated dose of 12.5 μl BCME/kg bw in peanut oil solution. All mice were killed and examined for lung tumours after 6 months. Control mice received a single s.c. dose of peanut oil, and a positive control group received a single s.c. dose of 1500 mg/kg bw urethane. In 50 males and 50 females injected with BCME, pulmonary tumours developed in 45% of the animals, with a multiplicity of 0.64 tumours per mouse. In addition, 1 mouse developed an injection site papilloma and another a fibrosarcoma; such tumours were not seen in control animals. In the vehicle controls, 7/50 mice had lung tumours, with a multiplicity of 0.14 tumours per mouse; and in the urethane group, 100% of the mice. had lung tumours, with a multiplicity of 17 tumours per mouse (Gargus et al., 1969).

Rat: A group of 20 female Sprague-Dawley rats received once-weekly s.c. injections of 3 mg BCME in 0.1 ml nujol (a refined pharmaceutical mineral oil with a low polycyclic aromatic hydrocarbon content). After 100 days, the dose was reduced to 1 mg per week because of local irritation at the initial dose level. Later (time not specified), the number of injections was reduced to 3 times per month, and was discontinued altogether after 300 days from the start of treatment. After a median survival time of 325 days, 2 local fibromas and 5 fibrosarcomas were seen. The first tumour was noted at 58 days. No increase in the incidence of remote tumours was seen. Control animals without treatment or injected repeatedly with 0.1 ml nujol exhibited no local tumours (Van Duuren et al., 1969).

3.2 Other relevant biological data

No data are available to the Working Group.

3.3 Observations in man

It has been reported by Thiess et al. (1973) that a retrospective investigation of a small group of BCME workers exposed to the compound between 1956 and 1962 revealed 6 cases of lung cancer amongst 18 men employed in a testing laboratory. Five of the 6 men were moderate smokers, and 1 was a non-smoker. Two further cases of lung cancer were found amongst a group of 50 production workers. Five of the total 8 reported cases are stated to have been oat-celled carcinomas. The exposure period ranged from 6 to 9 years, and the latent period from first exposure to diagnosis was from 8 to 16 years.

BCME is also found as an impurity (1-7%) in the related chloromethyl methyl ether (CMME). In an investigation in a chemical plant where exposure to CMME was observed, 14 cases of lung cancer, mainly oat cell carcinomas, were detailed (Figueroa et al., 1973). Three of the 14 individuals were not smokers, and 1 additional subject smoked a pipe only (see monograph on CMME for further details).

4. Comments on Data Reported and Evaluation

4.1 Animal data

Bis(chloromethyl)ether (BCME) is carcinogenic to mice following inhalation, skin application and subcutaneous administration. In newborn mice it is carcinogenic after a single subcutaneous exposure. In the rat it is carcinogenic by inhalation and subcutaneous administration.

4.2 Human data

A high incidence of predominantly oat-celled carcinoma in a small population of laboratory workers exposed to BCME strongly suggests that exposure to this compound constitutes a serious human lung cancer hazard.

There is also epidemiological evidence to suggest that exposure to BCME may constitute a lung cancer risk amongst workers exposed to it as a contaminant in the manufacture of the related CMME (see separate monograph on CMME).

5. References

Abrams, I.M. & Benezra, L. (1967) Ion-exchange polymers. In: Mark, H.F., ed., Encyclopedia of Polymer Science & Technology, New York, Wiley Interscience, Vol. 7, p. 692

Collier, L. (1972) Determination of bis(chloromethyl)ether at the ppb level in air samples by high-resolution mass spectroscopy. Environm. Sci. Technol., 6, 930

Figueroa, W.G., Raszkowski, R. & Weiss, W. (1973) Lung cancer in chloromethyl methyl ether workers. New Engl. J. Med., 288, 1096

Gargus, J.L., Reese, W.H., Jr & Rutter, H.A. (1969) Induction of lung adenomas in newborn mice by bis(chloromethyl)ether. Toxicol. appl. Pharmacol., 15, 92

Laskin, S., Kuschner, M., Drew, R.T., Cappiello, V.P. & Nelson, N. (1971) Tumors of the respiratory tract induced by inhalation of bis(chloromethyl)ether. Arch. environm. Hlth, 23, 135

Leong, B.K.J., Macfarland, H.N. & Reese, W.H., Jr (1971) Induction of lung adenomas by chronic inhalation of bis(chloromethyl)ether. Arch. environm. Hlth, 22, 663

Preussmann, R., Schneider, H. & Epple, F. (1969) Untersuchungen zum Nachweis alkylierender Agentien. II. Der Nachweis verschiedener Klassen alkylierender Agentien mit einer Modifikation der Farbreaktion mit 4-(4-Nitro-benzyl)-pyridin (NBP). Arzneimittel-Forsch., 19, 1059

Sawicki, E. & Sawicki, C.R. (1969) Analysis of alkylating agents. Application to air pollution. Ann. N.Y. Acad. Sci., 163, 895

Summers, L. (1955) The alpha-haloalkyl ethers. Chem. Rev., 55, 301

Thiess, A.M., Hey, W. & Zeller, H. (1973) Zur Toxikologie von Dichlordimethyläther - Verdacht auf kanzerogene Wirkung auch beim Menschen. Zbl. Arbeitsmed., 23, 97

US Government (1973) Occupational safety and health standards. US Federal Register, 38, No. 85, 10929

Van Duuren, B.L., Goldschmidt, B.M., Katz, C., Langseth, L., Mercado, G. & Sivak, A. (1968) Alpha-haloethers: a new type of alkylating carcinogen. Arch. environm. Hlth, 16, 472

Van Duuren, B.L., Katz, C., Goldschmidt, M., Frenkel, K. & Sivak, A. (1972) Carcinogenicity of halo-ethers. II. Structure-activity relationships of analogs of bis(chloromethyl)ether. J. nat. Cancer Inst., 48, 1431

Van Duuren, B.L., Sivak, A., Goldschmidt, B.M., Katz, C. & Melchionne, S. (1969) Carcinogenicity of halo-ethers. J. nat. Cancer Inst., 43, 481

CHLOROMETHYL METHYL ETHER*

1. Chemical and Physical Data

1.1 Synonyms and trade names

Chem. Abstr. No.: 107-30-2

CMME; chloromethyl ether[1]; dimethylchloroether; methyl chloromethyl
ether

1.2 Chemical formula and molecular weight

$ClCH_2OCH_3$ C_2H_5Cl0 Mol. wt: 80.5

1.3 Chemical and physical properties of the pure substance

(a) Description: Colourless liquid

(b) Boiling-point: 59°C at 760 mm

(c) Density: d_4^{20} 1.0605

(d) Refractive index: n_D^{20} 1.3974

(e) Solubility: Miscible with ethanol and ether and many other organic solvents

(f) Stability: Decomposes in the presence of water and in hot ethanol

(g) Volatility: Highly volatile

1.4 Technical products and impurities

Commercial chloromethyl methyl ether (CMME) contains a minimum of 95%
of the active ingredient. CMME typically contains several percent of bis
(chloromethyl)ether as an impurity (Collier, 1972).

---

* Considered by the Working Group in Lyon, June 1973.

[1] This name is more properly used for bis(chloromethyl)ether.

2. Production, Use, Occurrence and Analysis

(a) Production and use[1]

CMME can be produced by the reaction of methyl alcohol, formaldehyde and anhydrous hydrogen chloride.

No information is available on world production of CMME. The number of US producers has dropped from three in 1969 to one in 1973. In 1967, one company was reported to be making CMME in the Federal Republic of Germany. It seems likely that CMME is produced as a chemical intermediate in other countries.

CMME, both in a purified form and as a component of chloromethylating reaction mixture which may contain gross amounts of bis(chloromethyl)ether, is used as an intermediate in the synthesis of chloromethylated compounds which are often subsequently converted to aminated compounds by reaction with organic amines. One such utilization, and probably the largest single use for CMME, is in the preparation of ion-exchange resins. A modified polystyrene resin is chloromethylated and then treated with a tertiary amine or with a polyamine. Other chloromethylated compounds of commercial significance which may involve CMME in their manufacture include chloromethyl diphenyl oxide, 1-chloromethyl-naphthalene, di(chloromethyl)diphenyl oxide, di(chloromethyl)toluene and dodecylbenzyl chloride. No data are available on the total production of these chemicals or on the amount, if any, produced from CMME, since they are largely used as intermediates.

(b) Occurrence

CMME has not been reported to occur as such in nature. Since it hydrolyzes readily, it would not be expected to remain as such for protracted periods in waste streams from plants where it is produced or used.

---

[1] Data from Chemical Information Services, Stanford Research Institute, USA.

CMME is listed together with 13 other chemicals in the US Federal Register (US Government, 1973) as being subject to an Emergency Temporary Standard on certain carcinogens under an order made by the Occupational Safety and Health Administration, Department of Labor on 26 April, 1973.

(c) Analysis

The determination of CMME in air has been reported: CMME is hydrolyzed, and the products are determined turbidimetrically and colorimetrically (Vinogradova, 1962). Several methods for the detection of alkylating agents have been described (Preussmann et al., 1969; Sawicki & Sawicki, 1969).

3. Biological Data Relevant to the Evaluation of Carcinogenic Risk to Man

3.1 Carcinogenicity and related studies in animals

Virtually all samples of CMME tested contained 1-7% BCME.

(a) Skin application

Mouse: A cutaneous application of 0.1 ml of a 2% solution of CMME in benzene 3 times per week for 325 days to groups of 20 female ICR/Ha Swiss mice, observed further for a total of 540 days, gave no evidence of local tumourigenicity. However, the compound was active as an initiator. Administration of a single dose of 0.1 mg CMME in 0.1 ml benzene solution, followed 14 days later by thrice-weekly applications of 0.025 mg mixed phorbol esters in 0.1 ml acetone, resulted in 7 mice developing papillomas, the first one of which was seen at 259 days and 4 of which progressed to squamous cell carcinomas. The median survival time was 496 days. At a higher dose level, 1.0 mg CMME in 0.1 ml benzene given once followed by the promoting treatment, 5 mice developed papillomas, the first appearing at 140 days, and 1 of which progressed to a carcinoma. The average survival time was 488 days. Controls given a single cutaneous application of CMME at either dose level, followed by no treatment or by acetone, developed

241

no tumours. Two of 20 mice treated with mixed phorbol esters alone
had papillomas, the first appearing at 322 days and the median survival time being greater than 450 days. In a positive control group
pretreated with 0.15 mg benzo(a)pyrene in 0.1 ml benzene followed by
promotion with mixed phorbol esters, 20/20 mice developed papillomas;
the first of these appeared at 70 days, and 7 progressed to carcinomas.
The median survival time was 439 days (Van Duuren et al., 1969, 1972).

(b) Inhalation and/or intratracheal administration

Mouse: A group of 50 male strain A/He mice was exposed to an at-
mosphere containing a concentration of 0.006 mg/1 (6 mg/m^3) CMME in
exposure chambers for 6 hours per day on 5 days per week during 21
weeks. A total of 25/50 animals at risk had lung tumours, with an
average of 1.5 tumours per animal. In a control group exposed to fil-
tered room air for 130 days and held for 28 weeks, 20/49 mice at risk
had lung tumours, with an average of 0.9 tumours per mouse. A positive control group exposed to an aerosol of urethane at a concentration of 0.5 mg/1 (500 mg/m^3) for 130 exposure days in 28 weeks yielded
46/49 mice with lung tumours, with a tumour multiplicity of 54. Gas-
liquid chromatographic analysis of the CMME used indicated that BCME
was the only impurity, increasing from 0.3 to 2.6% during the test
period (Leong et al., 1971).

(c) Subcutaneous and/or intramuscular administration

Mouse: A group of 30 female ICR/Ha Swiss mice was injected s.c.
once per week with 300 µg CMME dissolved in 0.05 ml nujol (a purified
pharmaceutical mineral oil with a low polycyclic aromatic hydrocarbon
content) for a total of 685 days. Sarcomas at the injection site were
seen in 10 mice, the first tumour appearing at 308 days and the median
survival time being 496 days. There were no distant tumours. A control group receiving nujol alone exhibited no local tumours (Van Duuren
et al., 1972).

Newborn mice: Male and female ICR Swiss mice received a single
s.c. injection of a pre-determined maximum tolerated dose of 125 µl/kg
bw CMME in a solution of peanut oil. All animals were killed at 6

242

months, and 17/99 mice at risk had pulmonary tumours, with a multiplicity of 0.21. In the group of control mice injected with peanut oil alone, 7/50 had lung tumours, with a multiplicity of 0.14. A positive control group of 50 mice injected once with 1500 mg/kg bw urethane showed a 100% incidence of lung tumours, with a multiplicity of 17 (Gargus et al., 1969).

Rat: In a group of 20 female Sprague-Dawley rats given weekly s.c. injections of 3 mg CMME in 0.1 ml nujol for 300 days, 1 rat developed a fibrosarcoma at the injection site. The median survival time was 478 days, and all animals were evaluated at 515 days. Control animals developed no tumours (Van Duuren et al., 1969, 1972).

3.2 Other relevant biological data

No data are available to the Working Group.

3.3 Observations in man

Commercial CMME usually contains as an impurity a percentage (generally 1-7%) of highly carcinogenic bis(chloromethyl)ether. Insufficient epidemiological evidence is available at present to separate the carcinogenic effects of the two compounds.

Figueroa et al. (1973) have described an investigation in a chemical manufacturing plant with approximately 2,000 employees on the east coast of the United States (and see Nelson, 1973). In a preliminary observation during a 5-year period when 111 CMME workers were studied, 4 cases of lung cancer were diagnosed. This represents an incidence 8 times that observed in a control group of 2,804 with a similar cigarette smoking history (74-78%). The difference, not withstanding the small number, was stated to be highly significant statistically. Evidence for the existence of a specific lung cancer risk was further supported by the retrospective identification of a total of 14 cases, aged 35-55 years, all of whom had been employed in the production of CMME. Three had never smoked, 1 had smoked a pipe only. Length of exposure to CMME ranged from 3 to 14 years in 13 of the cases. It is noteworthy and unusual that 12 cases of lung cancer were confirmed histologically as being of an oat cell type. The ages of the lung cancer cases were somewhat lower than expected.

4. Comments on Data Reported and Evaluation

4.1 Animal data

Chloromethyl methyl ether (CMME) is almost invariably contaminated by bis(chloromethyl)ether (BCME), and the latter may be responsible for at least part of the observed carcinogenic activity. Such contaminated CMME has been found to be carcinogenic on subcutaneous injection in the mouse and possibly to be an initiator for mouse skin tumours. Inhalation in mice and subcutaneous injection in the rat produced equivocal evidence of carcinogenic activity.

4.2 Human data

One study based on 4 cases of oat cell lung cancer observed amongst 111 workers exposed to CMME (and its associated BCME impurity), followed for 5 years, suggests an increased risk of lung cancer.

5. References

Collier, L. (1972) Determination of bis(chloromethyl)ether at the ppb level in air samples by high-resolution mass spectroscopy. Environm. Sci. Technol., 6, 930

Figueroa, W.G., Raszkowski, R. & Weiss, W. (1973) Lung cancer in chloromethyl methyl ether workers. New Engl. J. Med., 288, 1096

Gargus, J.L., Reese, W.H., Jr & Rutter, H.A. (1969) Induction of lung adenomas in newborn mice by bis(chloromethyl)ether. Toxicol. appl. Pharmacol., 15, 92

Leong, B.K.J., Macfarland, H.N. & Reese, W.H., Jr (1971) Induction of lung adenomas by chronic inhalation of bis(chloromethyl)ether. Arch. environm. Hlth, 22, 663

Nelson, N. (1973) Carcinogenicity of halo ethers. New Engl. J. Med., 288, 1123

Preussmann, R., Schneider, H. & Epple, F. (1969) Untersuchungen zum Nachweis alkylierender Agentien. II. Der Nachweis verschiedener Klassen alkylierender Agentien mit einer Modifikation der Farbreaktion mit 4-(4-Nitrobenzyl)-pyridin (NBP). Arzneimittel Forsch., 19, 1059

Sawicki, E. & Sawicki, C.R. (1969) Analysis of alkylating agents. Application to air pollution. Ann. N.Y. Acad. Sci., 163, 895

US Government (1973) Occupational safety and health standards. US Federal Register, 38, No. 85, 10929

Van Duuren, B.L., Katz, C., Goldschmidt, B.M., Frenkel, K. & Sivak, A. (1972) Carcinogenicity of halo-ethers. II. Structure-activity relationship of analogs of bis(chloromethyl)ether. J. nat. Cancer Inst., 48, 1431

Van Duuren, B.L., Sivak, A., Goldschmidt, B.M., Katz, C. & Melchionne, S. (1969) Carcinogenicity of halo-ethers. J. nat. Cancer Inst., 43, 481

Vinogradova, V.A. (1962) The determination of monochlorodimethyl ether in the air. Novoe v. Oblasti Sanit.-Khim. Analiza (Raboty po Prom.-Sanit. Khim.), p. 158

1,4-BUTANEDIOL DIMETHANESULPHONATE*

1. Chemical and Physical Data

1.1 Synonyms and trade names

Chem. Abstr. No.: 55-98-1

1,4-Bis(methanesulphonoxy)butane; 1,4-bis(methanesulphonyloxy)butane; 1,4-dimethanesulphonoxybutane; 1,4-dimethanesulphonoxyl-butane; 1,4-di(methanesulphonyloxy)butane; 1,4-dimethylsulphonoxy-butane; 1,4-dimethylsulphonyloxybutane; methanesulphonic acid, tetramethylene ester; tetramethylene bis(methanesulphonate); tetramethylene dimethanesulphonate

Busulfan; Busulphan, Mablin; Mielevcin; Misulban; Mitosan; Myeloleukon; Myelosan; Myleran; Sulphabutin; C.B.2041; G.T.41

1.2 Chemical formula and molecular weight

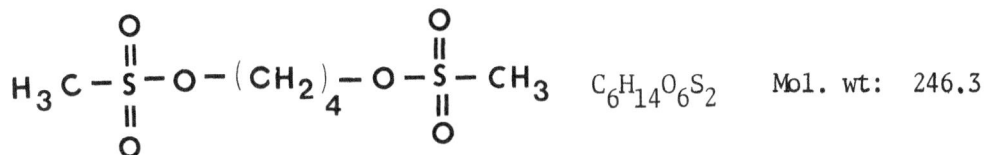

$$H_3C - \underset{\underset{O}{\overset{\|}{\|}}}{\overset{\overset{O}{\|}}{S}} - O - (CH_2)_4 - O - \underset{\underset{O}{\overset{\|}{\|}}}{\overset{\overset{O}{\|}}{S}} - CH_3 \qquad C_6H_{14}O_6S_2 \qquad Mol.\ wt:\ 246.3$$

1.3 Chemical and physical properties of the pure substance

(a) Description: White crystalline powder

(b) Melting-point: 114-118°C

(c) Solubility: 2.4 g in 100 ml acetone at 25°C, 0.1 g in 100 ml ethanol; almost insoluble in water

(d) Chemical reactivity: An active alkylating agent which hydrolyzes in water

---

* Considered by the Working Group in Lyon, June 1973.

1.4 Technical products and impurities

The USP grade product contains a minimum of 98% of 1,4-butanediol dimethanesulphonate (Myleran). Myleran is sold as a prescription drug in the form of tablets

2. Production, Use, Occurrence and Analysis

(a) Production and use[1]

According to the original patent on Myleran (Timmis, 1959) it can be produced by the reaction of 1,4-butanediol with methanesulphonyl chloride. One United States company, which has produced it since 1954, has a total annual production which is believed to be less than 500 kg.

This compound is used as a chemotherapeutic agent for the treatment of some forms of leukaemia, particularly chronic myelocytic leukaemia (Haddow & Timmis, 1953). Typical dosage level is 4-8 mg daily taken orally. Because of possible adverse side effects on the haematopoetic system, e.g., bleeding tendencies, decreased leucocyte count or depression of bone marrow activity, it is administered only when blood counts can be taken at least once weekly.

(b) Analysis

A procedure for the analysis of the pure drug and its tablet form is described (The Pharmacopoeia of the United States of America, 1970). General methods for the analysis of alkylating agents are applicable (Preussmann et al., 1969; Sawicki & Sawicki, 1969).

3. Biological Data Relevant to the Evaluation of Carcinogenic Risk to Man

3.1 Carcinogenicity and related studies in animals

(a) Oral administration

Rat: Schmähl & Osswald (1970) administered once-weekly doses of 0.13 mg/kg bw Myleran by gavage for 52 weeks to a group of 48 male ·

---

[1] Data from Chemical Information Services, Stanford Research Institute, USA.

BR46 rats. This dose is equivalent to the human therapeutic dose of 8 mg per day. Of 18 surviving animals observed for at least 26 months, 3 had tumours (1 subcutaneous fibrosarcoma, 1 squamous epithelioma of the lung and 1 benign thymoma). In 65/89 surviving controls, 3 rats had mammary sarcomas and 1 phaeochromocytoma. The incidence of both benign and malignant tumours was 11% in the controls and 17% in the treated animals. The mean induction time for tumours in controls was 23 ± 5 months and in the treated animals 22 ± 4 months.

(b) Intraperitoneal administration

Mouse: Groups of 30 female A mice received 0.5 mg Myleran per mouse on 3 consecutive days every 2 weeks over 7 weeks (total dose, 6 mg). At 18 weeks 6 animals were sacrificed, and 1 pulmonary tumour was found. At 30 weeks 21 mice were killed, and 8 were found to have pulmonary tumours. Of a control group of 30 mice examined at 18 weeks 3 had pulmonary tumours, and of a further group of 24 mice examined at 30 weeks 10 had pulmonary tumours (Shimkin, 1954).

(c) Other experimental systems

Intravenous injection: A high incidence of thymic lymphomas was reported in female RF mice by Conklin et al. (1965) after 4 i.v. doses, each of 12 mg/kg bw Myleran, given at 14-day intervals. Of 109 mice surviving after 30 days, 35% developed thymic lymphomas within an average latent period of 375 days, compared with 10% in 112 control mice. The incidence of ovarian tumours was also increased from 20% in the controls to 45% in the treated animals dying up to 621 days.

Random bred RF/Up male mice in groups ranging from 59 to 74 animals were given a single injection of 20 mg/kg bw Myleran in dimethylacetamide either before or after whole body irradiation (150 and 300 rads). Other groups received Myleran in the vehicle without irradiation, or vehicle only. A group of 314 mice served as untreated controls. Myleran administered in conjunction with X-rays increased the incidence of thymic lymphomas but lowered the incidence of irradiation-induced myeloid leukaemia (Upton et al., 1961).

3.2 Other relevant biological data

(a) Animals

In the rat and mouse 50-60% of a single dose of Myleran-^{35}S (10 mg/kg bw) injected intraperitoneally in arachis oil was excreted within 24 to 48 hours, mainly as methane sulphonic acid; a small amount of unchanged Myleran and two unidentified components were present. In the rabbit, methane sulphonic acid was the only metabolite found in the urine (Fox et al., 1960).

After i.p. injections of 2:3-^{14}C-Myleran in the rat, rabbit and mouse, 60% of the urinary radioactivity was found to be in the form of the 3-hydroxy tetrahydrothiophene-1,1-dioxide, a sulphone. It is suggested that in vivo Myleran undergoes a reaction with cysteine or a cysteinyl moiety to form a cyclic sulphonium ion, which in turn undergoes cleavage to the tetrahydrothiophene, oxidation to the 1,1-dioxide and biological hydroxylation to the 3-hydroxy compound (Roberts & Warwick, 1961).

3.3 Observations in man

Several reports (Nelson & Andrews, 1964; Min & Györkey, 1968; Comhaire et al., 1972; Dahlgren et al., 1972) describe dysplasias and atypical cell formation in the lung and cervix resembling precancerous lesions among patients who have received Myleran treatment. Two cases of cancer (1 bronchiolar cell carcinoma in a 72-year-old man and 1 mammary cancer in a 57-year-old woman) have been reported, but a causal association of these cancers with Myleran therapy cannot be established (Min & Györkey, 1968; Nelson & Andrews, 1964).

4. Comments on Data Reported and Evaluation

4.1 Animal data

Administration of 1,4-butanediol dimethanesulphonate (Myleran) to the mouse by intraperitoneal injection and to the rat by oral administration did not significantly increase the incidence of tumours. Intravenous administration in the mouse significantly increased the incidence of

250

thymic lymphomas and ovarian tumours. Myleran in conjunction with X-rays further augmented the incidence of thymic lymphomas. The increased incidences of thymic lymphomas and ovarian tumours are difficult to assess with respect to the carcinogenicity of Myleran in the mouse.

4.2 Human data

Although there is evidence that histological and cytological changes are associated with Myleran therapy, there is no firm evidence of an increased cancer risk among those treated.

5. References

Comhaire, F., Van Hove, W., Van Ganse, W. & Van der Straeten, M. (1972) Busulphan and the lungs. Absence of lung function disturbance in patients treated with busulphan. Scand. J. resp. Dis., 53, 265

Conklin, J.W., Upton, A.C. & Christenberry, K.W.(1965) Further observations on late somatic effects of radiomimetic chemicals and X-rays in mice. Cancer Res., 25, 20

Dahlgren, S., Holm, G., Svanborg, N. & Watz, R. (1972) Clinical and morphological side-effects of busulphan (myleran) treatment. Acta med. scand., 192, 129

Fox, B.W., Craig, A.W. & Jackson, H. (1960) The comparative metabolism of myleran-^{35}S in the rat, mouse and rabbit. Biochem. Pharmacol., 5, 27

Haddow, A. & Timmis, G.M. (1953) Mylran in chronic myeloid leukaemia. Chemical constitution and biological action. Lancet, i, 207

Min, K.-W. & Györkey, F. (1968) Interstitial pulmonary fibrosis, atypical epithelial changes and bronchiolar cell carcinoma following busulfan therapy. Cancer (Philad.), 22, 1027

Nelson, B.M. & Andrews, G.A. (1964) Breast cancer and cytologic dysplasia in many organs after busulphan (myleran). Am. J. clin. Path., 42, 37

The Pharmacopoeia of the United States of America (1970) 18th Revision, Bethesda, Maryland, Board of Trustees, p. 84

Preussmann, R., Schneider, H. & Epple, F. (1969) Untersuchungen zum Nachweis alkylierender Agentien. II. Der Nachweis verschiedener Klassen alkylierender Agentien mit einer Modifikation der Farbreaktion mix 4-(4-Nitrobenzyl)-pyridin (NBP). Arzneimittel-Forsch., 19, 1059

Roberts, J.J. & Warwick, G.P. (1961) The mode of action of alkylating
    agents. III. The formation of 3-hydroxytetrahydrothiophene-1,1-di-
    oxide from 1:4-dimethanesulphonyloxybutane (myleran), S-β-L-alanyl-
    tetrahydrothiophenium mesylate, tetrahydrothiophene and tetrahydro-
    thiophene-1,1-dioxide in the rat, rabbit and mouse. Biochem.
    Pharmacol., 6, 217

Sawicki, E. & Sawicki, C.R. (1969) Analysis of alkylating agents.
    Application to air pollution. Ann. N.Y. Acad. Sci., 163, 895

Schmähl, D. & Osswald, H. (1970) Experimentelle Untersuchungen über
    carcinogene Wirkungen von Krebs-Chemotherapeutica und Immuno-
    suppressiva. Arzneimittel-Forsch., 20, 1461

Shimkin, M.B. (1954) Pulmonary-tumor induction in mice with chemical
    agents used in chemical management of lymphomas. Cancer (Philad.),
    7, 410

Timmis, G.L. (December 15, 1959) Leukemia treatment, US Patent 2,917,432

Upton, A.C., Wolff, F.F. & Sniffen, E.P. (1961) Leukemogenic effect of
    myleran on the mouse thymus. Proc. Soc. exp. Biol. (N.Y.), 108, 464

1,3-PROPANE SULTONE*

1. Chemical and Physical Data

1.1 Synonyms and trade names

Chem. Abstr. No.: 1633-83-6

3-Hydroxy-1-propanesulphonic acid sultone; 1,2-oxathiolane-2,2-dioxide

1.2 Chemical formula and molecular weight

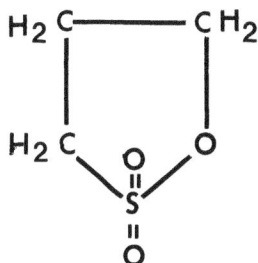

$C_3H_6O_3S$ Mol. wt: 122.1

1.3 Chemical and physical properties of the pure substance

(a) Description: Colourless liquid or white crystals

(b) Boiling-point: $112^{\circ}C$

(c) Melting-point: $31^{\circ}C$

(d) Density: d_4^{40} 1.393

(e) Refractive index: n_{40}^{D} 1.450

(f) Solubility: Readily soluble in many organic solvents such as ketones, esters and aromatic hydrocarbons; soluble in water (100g/l); insoluble in aliphatic hydrocarbons

(g) Stability: Half-life at $37^{\circ}C$ in phosphate buffer, pH 7.4, is 110 min. Hydrolyzes to 3-hydroxy-1-propanesulphonic acid.

---

* Considered by the Working Group in Lyon, June 1973.

1.4 Technical products and impurities

1,3-Propane sultone is available in a commercial grade which typically contains 99% active ingredient, 0.2% water and 0.8% acid (as 3-hydroxy-1-propanesulphonic acid).

2. Production, Use, Occurrence and Analysis

Two review articles on the properties and uses of 1,3-propane sultone have been published (Brémond, 1964; Fischer, 1964).

(a) Production and use[1]

1,3-Propane sultone is produced commercially by dehydrating gamma-hydroxy-propanesulphonic acid, which is prepared from sodium hydroxypropanesulphonate. This sodium salt is prepared by the addition of sodium bisulphite to allyl alcohol.

The only United States producer (which first produced semi-commercial quantities in 1963) is believed to have an annual production of less than 500 kg.

1,3-Propane sultone is used as a chemical intermediate to introduce the sulphopropyl group ($-CH_2CH_2CH_2SO_3-$) into molecules, and to confer water solubility and an anioinic character. A large number of sulphopropylated products are described and areas of possible utility are given in a review article (Fischer, 1964). Among the products and their potential uses mentioned in the review were the following: (i) derivatives of amines, alcohols, phenols, mercaptans, sulphides and amides useful as detergents, wetting agents, lathering agents and bacteriostats: (ii) soluble starches used in the textile industry; (iii) solubilized cellulose, which was reported to have soil-suspending properties; (iv) dyes; (v) an antistatic additive for polyamide fibres; (vi) cation-exchange resins (prepared by condensing the sulphonic acid product derived from phenol and propane sultone with formaldehyde); and (vii) phosphorus-containing sulphonic acids (produced from organic phosphines, neutral esters of trivalent phosphorous acids, and phosphorous and phosphoric triamides), useful as insecticides, fungicides, surfactants and vulcanization accelerators.

---

[1] Data from Chemical Information Services, Stanford Research Institute, USA.

(b) Occurrence

Propane sultone is not known to occur as such in nature. It may occur in the waste streams of plants making or using it, but because it is readily hydrolyzed it would not be expercted to remain as such for protracted periods of time.

(c) Analysis

No methods specific for the detection of 1,3-propane sultone have been published. Several methods for the detection of alkylating agents are available (Preussmann et al., 1969; Sawicki & Sawicki, 1969).

3. Biological Data Relevant to the Evaluation of Carcinogenic Risk to Man

3.1 Carcinogenicity and related studies in animals

(a) Oral administration

Rat: Druckrey et al. (1970) gave weekly doses of 30 mg/kg bw 1,3-propane sultone as a 3% aqueous solution by gavage to 12 rats, and 4/10 of the survivors developed malignant tumours between days 248 and 377. The tumours were: 1 glial-mesodermal mixed tumour and 1 adventitial cell sarcoma of the brain, 1 nephroblastoma and 1 subcutaneous spindle cell sarcoma. Ulland et al. (1971) showed that an aqueous solution of 1,3-propane sultone given by gavage twice weekly at doses of 56 mg/kg bw for 32 weeks and 28 mg/kg bw for 60 weeks produced an incidence of gliomas of 29/52 and 27/52, respectively, in rats killed at 60 weeks. Groups of 26 male and 26 female Charles River CD rats were used, including a normal and a positive control. The incidence of gliomas was similar at both dose levels in both sexes. In addition, several rats had leukaemia, ear duct tumours and adenocarcinomas of the small intestine. In 12 control rats killed at 61 weeks, only 1 pituitary adenoma was found, although 1 female control died after 33 weeks with a cerebral glioma.

(b) Subcutaneous and/or intramuscular administration

Mouse: A group of 30 female ICR/Ha Swiss mice was given weekly s.c. injections of 0.3 mg 1,3-propane sultone in 0.05 ml distilled water. This produced tumours at the injection site in 21/30 mice within 63 weeks (1 papilloma, 7 adeno-acanthomas, 12 sarcomas and 1 undifferentiated carcinoma). No tumours were seen in 30 controls after 78 weeks (Van Duuren et al., 1971).

Rat: In two groups of 12 BD rats given weekly s.c. doses of 15 and 30 mg/kg bw 1,3-propane sultone as a 1% solution in arachis oil, 7/12 and 11/11 rats, respectively, died with local sarcomas (myosarcomas and fibrosarcomas) at the site of injection within 217 to 360 days (total dose, up to 390 mg/kg bw). An adenocarcinoma of the ileum was seen in 1 rat given the lower dosage (Druckrey et al., 1968, 1970).

Weekly s.c. injections of 15 mg/kg bw in water (total dose, 225 mg/kg bw) resulted in local sarcomas in all of 18 treated rats dying between 220 and 343 days. A single s.c. dose of 100 mg/kg produced local sarcomas in all of 18 treated rats dying between 208 and 387 days. In addition, 1 rat developed a malignant neurocytoma. Of 18 rats given a single s.c. injection of 30 mg/kg bw, 12 developed sarcomas at the injection site within 400 days. Of 20 rats given 10 mg/kg bw s.c., 4/15 survivors developed local sarcomas within 500 days (Druckrey et al., 1970).

(c) Other experimental systems

Intravenous administration: Druckrey et al. (1970) administered weekly injections of a 2% solution of 1,3-propane sultone in water to 10 BD rats at a dose of 40 mg/kg bw. Treatment was stopped after 16 weeks (total dose, 560 mg/kg bw) due to sclerosis of the tail vein. In one rat dying after 280 days, a sarcoma of the mediastinum with metastases in the right lung and kidney were observed. In two rats dying after 398 and 410 days, 1 glial-mesodermal mixed tumour of the brain and 1 neurosarcoma of the plexus cervicalis were observed.

In 3/8 rats surviving 381 to 492 days after weekly injections of 20/mg kg bw (total dose, 570 mg/kg bw), 1 nephroblastoma, 1 carcinoma of the ileocaecal region and 1 mixed glial-mesodermal tumour of the brain together with a mammary carcinoma were observed. In 2/12 rats given 10 mg/kg bw/week (total dose, 300 mg/kg bw), 1 ganglioneuroma of the adrenal gland and 1 neurocytoma of the trigeminus nerve developed after 296 and 469 days.

Of 32 rats given a single dose of 150 mg/kg bw, 1 died after 235 days with a glioma of the brain, and 9 died with malignant tumours at a variety of sites within 459 days.

Pre-natal exposure: A single i.v. injection of 20 mg/kg bw 1,3-propane sultone given to pregnant rats on the 15th day of gestation produced malignant neurogenic tumours in 3/25 offspring; a dose of 60 mg/kg bw produced malignant tumours in 4/14 offspring, including 2 neurogenic tumours, 1 tumour of the pancreas and 1 of the ovary (Druckrey et al., 1970).

3.2 Other relevant biological data

No data are available to the Working Group.

3.3 Observations in man

No data are available to the Working Group.

4. Comments on Data Reported and Evaluation[1]

4.1 Animal data

1,3-Propane sultone has a carcinogenic effect in the rat when given orally, intravenously or by pre-natal exposure, and a local carcinogenic effect in the mouse and the rat when given subcutaneously. It is carcinogenic in the rat after single-dose exposures.

4.2 Human data

No epidemiological data are available to the Working Group.

---

[1] See also the section "Extrapolation from animals to man" in the introduction to this volume.

5. References

Brémond, J. (1964) Propane sultone, propriétés et emplois. Rév. Prod. chim., 15 Sept., 433

Druckrey, H., Kruse, H. & Preussmann, R. (1968) Propane sultone, a potent carcinogen. Naturwissenschaften, 55, 449

Druckrey, H., Kruse, H., Preussmann, R., Ivankovic, S., Landschütz, Ch. & Gimmy, J. (1970) Cancerogene alkylierende Substanzen. IV. 1,3-Propanesulton und 1,4-Butansulton. Z. Krebsforsch., 75, 69

Fischer, R.F. (1964) Propane-sultone. Int. engng Chem., 56, 41

Preussmann, R., Schneider, H. & Epple, F. (1969) Untersuchungen zum Nachweis alkylierender Agentien. II. Der Nachweis verschiedener Klassen alkylierender Agentien mit einer Modifikation der Farbreaktion mit 4-(4-Nitrobenzyl)-pyridin (NBP). Arzneimittel-Forsch., 19, 1059

Sawicki, E. & Sawicki, C.R. (1969) Analysis of alkylating agents. Application to air pollution. Ann. N.Y. Acad. Sci., 163, 895

Ulland, B., Finkelstein, M., Weisburger, E.K., Rice, J.M. & Weisburger, J.H. (1971) Carcinogenicity of the industrial chemicals propylene imine and propane sultone. Nature (Lond.), 230, 460

Van Duuren, B.L., Melchionne, S., Blair, R., Goldschmidt, B.M. & Katz, C. (1971) Carcinogenicity of isoesters of epoxides and lactones: Aziridine ethanol, propane sultone and related compounds. J. nat. Cancer Inst., 46, 143

β-PROPIOLACTONE*

1. Chemical and Physical Data

1.1 Synonyms and trade names

Chem. Abstr. No.: 57-57-8

Hydracrylic acid, β-lactone; 3-hydroxypropionic acid, β-lactone; 2-oxetanone; propanolide; propiolactone; propionic acid 3-hydroxy-β-lactone; β-propionolactone; β-proprolactone

Betaprone; BPL

1.2 Chemical formula and molecular weight

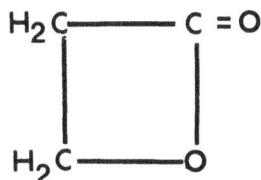

H_2C————$C = O$

H_2C————O

$C_3H_4O_2$ Mol. wt: 72.1

1.3 Chemical and physical properties of the pure substance

(a) Description: Colourless liquid, with a slightly sweetish odour

(b) Boiling-point: $162^{\circ}C$ (decomposition); $51^{\circ}C$ at 10 mm

(c) Melting-point: $-33.4^{\circ}C$

(d) Density: D_4^{20} 1.1460

(e) Refractive index: n_D^{20} 1.14131

(f) Solubility: At $25^{\circ}C$ it is soluble in water (37% v/v) with hydrolysis. Miscible with ethanol (reacts), with acetone, ether and chloroform and probably with most polar organic solvents and lipids.

(g) Volatility: Vapour pressure at $25^{\circ}C$ is 3.4 mm Hg

---

* Considered by the Working Group in Lyon, June 1973.

(h) <u>Stability</u>: Tends to polymerize on storage (Hoffman & Warshowsky, 1958); should be stored at low temperatures ($0^{o}C$)

(i) <u>Chemical reactivity</u>: Very high chemical reactivity due to the presence of a strained four-membered lactone ring. It has a half-life of about three hours in water at $25^{o}C$ and hydrolyzes to 3-hydroxy-propionic acid. It is a nucleophilic alkylating agent which reacts readily, for example, with acetate, halogen, thiocyanate, thiosulphate, hydroxyl and sulphydryl ions. It reacts with amino acids (Taubman & Atassi, 1968). With cysteine, it forms S-2-carboxyethylcysteine (Dickens & Jones, 1961).

1.4 Technical products and impurities

β-Propiolactone is available in a grade containing 97% minimum active ingredient, 0.2% maximum water content and 1.0% maximum acrylic acid content. One source reported that the impurities in fresh and aged commercial samples include acrylic acid, acrylic anhydride, acetic acid and acetic anhydride in amounts of less than 1% (Palmes et al., 1962).

2. Production, Use, Occurrence and Analysis

(a) Production and use[1]

β-Propiolactone has been produced commercially in the United States since 1958 by the reaction of ketene with formaldehyde. Since only one company produced this chemical, no US production quantities are published. It was recently reported that this company will close its plant by the end of 1973 (Hyatt, 1973). Data are published on US imports of all monolactones other than butyrolactone, and of lactones used in perfumery. These imports amounted to 160 thousand kg (two-thirds of this was from the Federal Republic of Germany), but the amount of β-propiolactone included is not known.

In 1967, one company was reported to be producing β-propiolactone in the Federal Republic of Germany. In 1970, β-propiolactone was produced by

---

[1] Data from Chemical Information Services, Stanford Research Institute, USA.

one manufacturer in Canada. The production capacity of these companies is not known.

β-Propiolactone is believed to be used mainly as an intermediate in the production of acrylic acid and esters. At the present time, two US companies have a combined capacity of approximately 40 million kg per year of acrylic acid and esters based on β-propiolactone.

β-Propiolactone has been reported as being used in the sterilization of blood plasma (Hartman et al., 1954; Hartman & LoGrippo, 1957), vaccines (Hartman & LoGrippo, 1957; Orlans & Jones, 1958; Keeble & Heymann, 1965; Soltys, 1967; Perlman & Malley, 1969), tissue grafts (Rains et al., 1956; Hartman & LoGrippo, 1957), surgical instruments (Allen & Murphy, 1960) and enzymes (Stokes, 1971) and as a vapour phase disinfectant in enclosed spaces (Hoffman & Warshowsky, 1958; Bruch, 1961). Such uses may now have been discontinued in several countries.

(b) Occurrence

β-Propiolactone has not been reported to occur as such in nature. Although it may be present in the waste streams from plants where it is made or used, it has a short half-life in water.

This compound is listed with 13 other chemicals in the US Federal Register (US Government, 1973) as being subject to an Emergency Temporary Standard on certain carcinogens under an order made by the Occupational Safety and Health Administration, Department of Labor on April 26, 1973.

(c) Analysis

Schmitz-Masse (1972) describes the identification and quantitative determination of β-propiolactone by gas-liquid chromatography. It can also be determined by polarography after conversion to β-nitropropionic acid with sodium nitrite (Pellerin & Letavernier, 1971). Several methods for the detection of alkylating agents are available (Preussmann et al., 1969; Sawicki & Sawicki, 1969).

3. Biological Data Relevant to the Evaluation of Carcinogenic Risk to Man

3.1 Carcinogenicity and related studies in animals

(a) Oral administration

Rat: Weekly doses of 10 mg β-propiolactone in 0.5 ml tricaprylin given to female Sprague-Dawley rats for 487 days produced squamous cell carcinomas of the forestomach in 3/5 rats. No such tumours were seen in 5 controls given weekly doses of 0.5 ml tricaprylin (Van Duuren et al., 1966).

(b) Skin application

Mouse: Weekly painting of 0.3 ml of a 2.5% solution of β-propiolactone in acetone for 52 weeks to 10 stock "S" mice produced papillomas (the first appearing at 27 weeks) in 5/9 mice surviving for 55 weeks. Two tumours underwent a malignant change to squamous cell carcinomas by 40 weeks, and tumours in 2 other mice were regarded as "probably malignant" (Roe & Glendenning, 1956). In another experiment 25 stock "S" mice were given 5 weekly applications of a 5% or 10% solution to produce skin irritation, followed by weekly applications of a 2.5% solution for 35 weeks. In this study scarring appeared to favour the early appearance of tumours, 3 malignant tumours appearing by 21 weeks in 1 mouse in which scarring persisted (Roe & Glendenning, 1956).

Four groups, each of 30 male Swiss mice, were given thrice-weekly applications of 100 mg β-propiolactone as a 0.25, 0.8, 2.5 or 5% solution in acetone for life. Two further groups of 30 mice received doses of 0.8% or 2.5% β-propiolactone in corn oil thrice weekly for 5 weeks and once weekly thereafter. The control groups received acetone (90 mice) or corn oil (80 mice) alone. Papillomas developed in 12, 15, 18 and 21 mice in each of the first 4 groups; in 27 and 15 of the next 2 groups; and in 11/90 and 14/80 mice of the 2 control groups. Malignant tumours developed in 3, 2, 9, 11, 12, 3, 1 and 0 mice in these 8 groups. Most of the animals in the first 5 groups died by 300 days. In another group of mice given doses of 0.3 ml of a 2.5% solution in acetone once weekly, papillomas developed in 6/29 mice and malignant tumours in 2 cases; all mice died before 300 days (Palmes et al., 1962). Short-term application in the same study also produced papillomas from which some malignant tumours developed. Similar results were obtained by Searle (1961).

When stock strain "S" mice were treated with β-propiolactone under a variety of protocols, subsequent treatment with croton oil demonstrated a cocarcinogenic effect (Roe & Salaman, 1955).

Hamster: Twice-weekly applications of 0.5 ml of a 2.5% solution of β-propiolactone in acetone to 17 male golden hamsters resulted in the development of papillomas in 4/13, melanomas in 4/13 (weeks 28 to 35), keratoacanthomas in 4/13 and squamous cell carcinomas in the skin of 2/13 hamsters surviving 32 to 100 weeks. No controls are described (Parish & Searle, 1966b).

Guinea-pig: Five male and 4 female guinea pigs received twice-weekly applications of 0.5 ml per site of a 2.5% solution of β-propiolactone in acetone to 2 sites and of a 5% solution in acetone to 2 other sites for several months, after which all 4 sites received the 5% solution. In 3/7 animals surviving for 85 to 168 weeks keratoacanthomas developed, and 1 additional animal developed a melanoma (Parish & Searle, 1966a). The significance of these results is questionable since no controls were included in the study.

(c) Subcutaneous and/or intramuscular administration

Mouse: Twice-weekly s.c. injections of 20 μg β-propiolactone in 0.1 ml arachis oil into mice of the Tuck No. 1 strain over 81 weeks resulted in the development of local tumours in 10/20 animals surviving the appearance of the first tumour at 43 weeks. Controls given arachis oil over 72 weeks exhibited 1 mammary adenoma among 19 survivors (Dickens & Jones, 1965).

Following weekly s.c. injections of 0.73 mg β-propiolactone in 0.05 ml tricaprylin for 503 days to 30 female ICR/Ha Swiss mice, 3 mice developed squamous papillomas at the injection site, and 18 mice developed malignant tumours (9 fibrosarcomas, 3 adenocarcinomas and 6 squamous cell carcinomas) at the injection site. No local tumours developed in 110 controls given weekly injections of 0.05 ml tricaprylin over 532 to 581 days (Van Duuren et al., 1966).

Rat: Following twice-weekly s.c. injections of β-propiolactone for 90 days, totalling 440 mg/kg bw in 6 males and 480 mg/kg bw in 6 females dosed in 22 and 24 ml/kg bw arachis oil, respectively, local sarcomas developed in 5/6 males and in 4/6 females within 192 to 386 days. Experiments in various arachis oil control groups were run in parallel. Local sarcomas appeared in 13/48 males and females given twice-weekly injections of arachis oil, totalling 70-397 ml/kg bw over 97 to 300 days and observed for up to 2 years (Walpole et al., 1954).

Twice-weekly s.c. doses of 2 mg β-propiolactone in 0.5 ml sesame oil for 38 weeks produced local sarcomas in 10/12 rats; 12 controls injected with oil alone had no tumours (Dickens et al., 1956).

Following twice-weekly s.c. injections of 0.1 or 1 mg β-propiolactone in arachis oil or 2 mg in water in groups of 10 male Wistar rats for 34, 44 and 33 weeks, respectively, injection-site sarcomas developed in 4/4, 10/10 and 2/4 rats surviving the onset of the first tumour, i.e., at 25 to 31 weeks. Animals were observed up to 55 weeks. No local sarcomas developed in 6 controls given repeated injections of 0.5 ml arachis oil for 54 weeks (Dickens & Jones, 1961).

After several weekly doses of 33 mg in 0.1 ml tricaprylin, the dose was reduced to 11 mg per injection due to skin irritation, and finally, after 93 days, to 4 mg per injection for the remainder of the test (total duration, 378 days). Local sarcomas developed in 13/20 female Sprague-Dawley rats. No local tumours developed in 40 controls given injections of 0.1 ml tricaprylin over 555 days (Van Duuren et al., 1967).

(e) Intraperitoneal administration

Mouse: A single dose of 0.1 mg/kg bw β-propiolactone given in oil to 35 male and 33 female B6AF$_1$ mice, 9 to 11 days after birth, produced lymphomas in 8.8% of suckling males and 20% of suckling females, compared with 0% in both male and female controls given olive oil alone. Hepatomas developed in 22/34 males and 0/30 females, compared with 1/25 and 0/25 in olive oil-treated controls. The hepatomas did not produce metastases. The incidence of hepatomas was not significantly increased after a single i.p. dose of 80 mg/kg bw given to 22 male and 24 female adult B6AF$_1$ mice (Chernozemski & Warwick, 1970).

3.2 Other relevant biological data

(a) Animals

β-Propiolactone binds in vivo to DNA, RNA and proteins of mouse skin. The degree of tumour-initiating activity is proportional to the extent of DNA binding but not to the extent of RNA or protein binding. A major RNA and DNA binding product is 7-(2-carboxyethyl) guanine. S-2-Carboxyethylcysteine was found in the acid hydrolysate of protein isolated from the skin of β-propiolactone-treated mice (Boutwell et al., 1969). Two weeks after the application of 3H-β-propiolactone to mouse skin, the radioactivity bound to skin DNA was less than 6% of the two-hour value (Colburn & Boutwell, 1966).

β-Propiolactone inhibits DNA synthesis in vivo in mouse skin (Hennings & Boutwell, 1969), and is claimed to result in an association of nucleic acids and cellular membranes, as demonstrated by CsCl gradient fractionation (Kubinski et al., 1972).

(b) Carcinogenicity of metabolites

β-Hydroxypropionic acid, the hydrolysis product of β-propiolactone, failed to produce either local sarcomas in a subcutaneous study in rats (Dickens & Jones, 1963) or skin tumours after applications to the skin in mice (Searle, 1961).

β-Propiolactone can react with chloride ion to form 3-chloropropionic acid, especially in blood plasma (Searle, 1961). This compound failed to show tumour-initiating activity in a mouse skin-painting study lasting 30 weeks (Colburn & Boutwell, 1968). Also, structurally-related 3-bromopropionic acid gave a negative result in a mouse skin-painting test lasting 40 weeks (Searle, 1961).

3.3 Observations in man

No data are available to the Working Group.

4. Comments on Data Reported and Evaluation[1]

4.1 Animal data

β-Propiolactone is carcinogenic in the mouse by skin application, sub-cutaneous or intraperitoneal injection, and in the rat by subcutaneous injection. Oral administration in the rat gave some indication of carcinogenic activity. The results obtained in the hamster and guinea pig are equivocal. It is carcinogenic to mice after a single-dose exposure.

4.2 Human data

No epidemiological data are available to the Working Group.

---

[1] See also the section "Extrapolation from animals to man" in the introduction to this volume.

5. References

Allen, H.F. & Murphy, J.T. (1960) Sterilization of instruments and materials with β-propiolactone. J. amer. Med. Ass., 172, 1759

Boutwell, R.K., Colburn, N.H. & Muckerman, C.C. (1969) In vivo reactions of β-propiolactone. Ann. N.Y. Acad. Sci., 163, 751

Bruch, C.W. (1961) Decontamination of enclosed spaces with β-propiolactone vapor. Amer. J. Hyg., 73, 1

Chernozemski, I.M. & Warwick, G.P. (1970) Production of hepatomas in suckling mice after single application of β-propiolactone. J. nat. Cancer Inst., 45, 709

Colburn, N.H. & Boutwell, R.K. (1966) The binding of β-propiolactone to mouse skin DNA in vivo; its correlation with tumor-initiating activity. Cancer Res., 26, 1701

Colburn, N.H. & Boutwell, R.K. (1968) The binding of β-propiolactone and some related alkylating agents to DNA, RNA, and protein of mouse skin; relation between tumor-initiating power of alkylating agents and their binding to DNA. Cancer Res., 28, 653

Dickens, F. & Jones, H.E.H. (1961) Carcinogenic activity of a series of reactive lactones and related substances. Brit. J. Cancer, 15, 85

Dickens, F. & Jones, H.E.H. (1963) Further studies on the carcinogenic and growth-inhibitory activity of lactones and related substances. Brit. J. Cancer, 17, 100

Dickens, F. & Jones, H.E.H. (1965) Further studies on the carcinogenic action of certain lactones and related substances in the rat and mouse. Brit. J. Cancer, 19, 392

Dickens, F., Jones, H.E.H. & Williamson, D.H. (1956) Carcinogenesis by simple chemical agents. A.R. Brit. Emp. Cancer Campgn, 34, 100

Hartman, F.W. & LoGrippo, G.A. (1957) β-Propiolactone in sterilization of vaccines, tissue grafts and plasma. J. amer. ed. Ass., 164, 258

Hartman, F.W., LoGrippo, G.A. & Kelly, A.R. (1954) Preparation and sterilization of blood plasma. Amer. J. clin. Path., 24, 399

Hennings, H. & Boutwell, R.K. (1969) The inhibition of DNA synthesis by initiators of mouse skin tumorigenesis. Cancer Res., 29, 510

Hoffman, R.K. & Warshowsky, B. (1958) β-Propiolactone vapor as a disinfectant. Appl. Microbiol., 6, 358

Hyatt, J.C.   (1973)   US mulls rules for handling of chemicals that can lead to cancer in plant workers.   Wall Street J., April 11, 36

Keeble, S.A. & Heymann, C.S.   (1965)   Effect of β-propiolactone and thiomersal on growth of B.H.K. cells in the hamster cheek pouch.   Nature (Lond.), 208, 1125

Kubinski, H., Andersen, P.R. & Kellicutt, L.M.   (1972)   Association of nucleic acids and cellular membranes induced by a carcinogen, β-propiolactone.   Chem.-biol. Interactions, 5, 279

Orlans, E.S. & Jones, V.E.   (1958)   β-Propiolactone as a toxoiding agent.   Nature (Lond.), 182, 1216

Palmes, E.D., Orris, L. & Nelson, N.   (1962)   Skin irritation and skin tumor production by β-propiolactone (BPL).   Amer. industr. Hyg. Ass. J., 23, 257

Parish, D.J. & Searle, C.E.   (1966a)   The carcinogenicity of β-propiolactone and 4-nitroquinoline N-oxide for the skin of the guinea-pig.   Brit. J. Cancer, 20, 200

Parish, D.J. & Searle, C.E.   (1966b)   The carcinogenicity of β-propiolactone and 4-nitroquinoline N-oxide for the skin of the golden hamster.   Brit. J. Cancer, 20, 206

Pellerin, F. & Letavernier, J.F.   (1971)   Détéction de la β-propiolactone par polarography.   Ann. pharm. franç., 29, 444

Perlman, F. & Malley, A.   (1969)   Effects of β-propiolactone (Betaprone) on reaginic (skin-sensitizing) antibodies.   J. Allergy, 44, 10

Preussmann, R., Schneider, H. & Epple, F.   (1969)   Untersuchungen zum Nachweis alkylierender Agentien.   II.   Der Nachweis vershiedener Klassen alkylierender Agentien mit einer Modifikation der Farbreaktion mit 4-(4-Nitrobenzyl)-pyridin (NBP).   Arzneimittel Forsch., 19, 1059

Rains, A.J.H., Crawford, N., Sharpe, S.H., Shrewsbury, J.F.D. & Barson, G.J.   (1956)   Management of an artery-graft bank with special reference to sterilisation by β-propiolactone.   Lancet, ii, 830

Roe, F.J.C. & Glendenning, O.M.   (1956)   The carcinogenicity of β-propiolactone for mouse-skin.   Brit. J. Cancer, 10, 357

Roe, F.J.C. & Salaman, M.H.   (1955)   Further studies on incomplete carcinogenesis: triethylene melamine (T.E.M.), 1,2-benzanthracene and β-propiolactone, as initiators of skin tumour formation in the mouse.   Brit. J. Cancer, 9, 177

Sawicki, E. & Sawicki, C.R.   (1969)   Analysis of alkylating agents.   Application to air pollution.   Ann. N.Y. Acad. Sci., 163, 895

Schmitz- Masse, M.O. (1972) Analyse de la β-propiolactone par chromatographie gaz-liquide. I. Récherche d'un résidu de β-propiolactone dans le vaccin antirabique Pasteur. II. Etude de la courbe d'hydrolyse de la β-propiolactone à pH 7.4. J. Chromat., 70, 128

Searle, C.E. (1961) Experiments on the carcinogenicity and reactivity of β-propiolactone. Brit. J. Cancer, 15, 804

Soltys, M.A. (1967) Comparative studies of immunogenic properties of Trypanosoma brucei inactivated with β-propiolactone and with some other inactivating agents. Canad. J. icrobiol., 13, 743

Stokes, K.J. (1971) β-Propiolactone as an agent for enzyme sterilization. J. clin. Path., 24, 658

Taubman, M.A. & Atassi, M.Z. (1968) Reaction of β-propiolactone with amino acids and its specificity for methionine. Biochem. J., 106, 829

US Government (1973) Occupational safety and health standards. US Federal Register, 38, No. 85, 10929

Van Duuren, B.L., Langseth, L., Orris, L., Baden, M. & Kuschner, M. (1967) Carcinogenicity of epoxides, lactones, and peroxy compounds. V. Subcutaneous injection in rats. J. nat. Cancer Inst., 39, 1213

Van Duuren, B.L., Langseth, L., Orris, L., Teebor, G., Nelson, N. & Kuschner, M. (1966) Carcinogenicity of epoxides, lactones, and peroxy compounds. IV. Tumor response in epithelial and connective tissue in mice and rats. J. nat. Cancer Inst., 37, 825

Walpole, A.L., Roberts, D.C., Rose, F.L., Hendry, J.A. & Homer, R.F. (1954) Cytotoxic agents. IV. The carcinogenic actions of some monofunctional ethyleneimine derivatives. Brit. J. Pharmacol., 9, 306

DIMETHYL SULPHATE*

1. Chemical and Physical Data

1.1 Synonyms and trade names

Chem. Abstr. No.: 77-78-1

Dimethyl monosulphate; methyl sulphate; sulphuric acid, dimethyl ester

1.2 Chemical formula and molecular weight

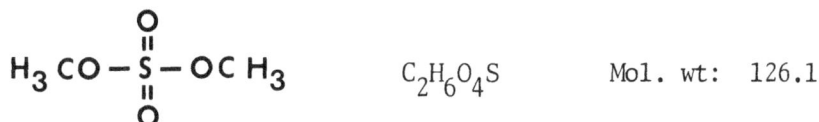

$$H_3CO - \overset{\overset{O}{\|}}{\underset{\underset{O}{\|}}{S}} - OCH_3 \qquad C_2H_6O_4S \qquad \text{Mol. wt: } 126.1$$

1.3 Chemical and physical properties of the pure substance

(a) Description: Colourless, oily liquid

(b) Boiling-point: 188°C (with decomposition); 76°C at 15 mm

(c) Melting-point: - 27°C

(d) Density: d_4^{20} 1.332

(e) Refractive index: n_D^{20} 1.3874

(f) Solubility: Miscible with many polar organic solvents and aromatic hydrocarbons, but sparingly soluble in carbon disulphide and aliphatic hydrocarbons

(g) Volatility: Vapour pressure at room temperature is 0.1 mm Hg

(h) Stability: Stable at room temperature; hydrolysis in water is rapid

(i) Reactivity: An active alkylating agent

---

* Considered by the Working Group in Lyon, June 1973.

1.4 Technical products and impurities

Dimethyl sulphate is available as a technical grade product which contains small amounts of acid impurities.

2. Production, Use, Occurrence and Analysis

A review on sulphuric and sulphurous esters including dimethyl sulphate was published in 1969 (Fuchs, 1969).

(a) Production and use[1]

Dimethyl sulphate has been produced commercially for at least 50 years. It can be made by the continuous reaction of dimethyl ether with sulphur trioxide. No information is available on world commercial production. In 1967, two companies were reported to be manufacturing dimethyl sulphate in the Federal Republic of Germany. One producer was reported in Italy in 1969, and one producer was reported in the United Kingdom in 1970. One United States company manufactures dimethyl sulphate.

Dimethyl sulphate is used mainly as an alkylating agent for converting active-hydrogen compounds such as phenols, amines and thiols to the corresponding methyl derivatives. Since alternative methylating agents are available for the production of most methyl derivatives, it is not possible to establish whether or not they are produced with dimethyl sulphate. In the case of the quaternary ammonium methosulphate salts (produced by the reaction of dimethyl sulphate with tertiary amines), dimethyl sulphate must be used as the alkylating agent. US production of eight compounds of this type amounted to at least 500 kg each in 1970 (US Tariff Commission, 1972). Included in the group were the following six cationic surfactants: dimethyl dioctadecyl ammonium methosulphate; (3-lauramidopropyl) trimethyl ammonium methosulphate; (3-oleamidopropyl) trimethyl ammonium methosulphate; the methosulphate of a stearic acid-diethanolamine condensate; the methosul-

---

[1] Data from Chemical Information Services, Stanford Research Institute, USA.

phate of N-(2-hydroxethyl)-N,N',N'-tris(2-hydroxypropyl) ethylenediamine distearate; and the methosulphate of N,N,N',N'-tetrakis (2-hydroxypropyl) ethylenediamine dioleate. Also included were two anti-cholinergic agents: diphemanil methyl sulphate and hexocyclium methyl sulphate.

A parasympathomimetic agent, neostigmine methyl sulphate, is also produced in the US, but no information on production quantity is available; and in 1973, production in the US of another cationic surfactant, 2-hepta-decyl-1-methyl-1-(2-stearamidoethyl)-2-imidazolinium methosulphate, was reported.

(b) Occurrence

Dimethyl sulphate has not been reported to occur as such in nature. Although it could occur in the waste streams from plants where it is made or used, it hydrolyzes rapidly.

(c) Analysis

A simple colorimetric test, involving conversion to nitromethane and condensation with 1,2-naphthoquinone-4-sulphonic acid, has been published (Feigl & Goldstein, 1957). Several methods for the detection of alkyla-ting agents have been described (Preussmann et al., 1969; Sawicki & Sawicki, 1969).

3. Biological Data Relevant to the Evaluation
of Carcinogenic Risk to Man

3.1 Carcinogenicity and related studies in animals

(a) Inhalation and/or intratracheal administration

Rat: Of 27 BD rats exposed for 1 hour to 10 ppm (55 mg/m^3) di-methyl sulphate 5 times weekly during 1 hour for 130 days, 5/15 ani-mals surviving 643 days developed malignant tumours, including 3 squamous carcinomas of the nasal cavity, 1 mixed tumour of the cere-bellum and 1 lymphosarcoma of the thorax with multiple metastases to the lungs. Of 20 rats exposed to 3 ppm (17 mg/m^3) 3 died with tu-mours: 1 with a neurocytoma, 1 with an aesthesioneuroepithelioma of

the olfactory nerve and 1 with a squamous carcinoma of the nasal cavity. No data on control animals were reported (Druckrey et al.,
1970).

(b) Subcutaneous and/or intra-muscular administration

Rat: Among 8 BD rats given weekly s.c. doses of 16 mg/kg bw
dimethyl sulphate in oil (total dose, 784 mg/kg bw) 4/6 survivors
developed local sarcomas within 236 to 414 days, 1 with metastases
to the lungs. Among 12 BD rats receiving 8 mg/kg bw in oil weekly
for 394 days (total dose, 466 mg/kg bw) 7/11 survivors showed local
sarcomas, 3 of which had metastasized to the lungs and regional lymph
nodes. The mean tumour induction time was 500 days. One rat in this
group developed a liver carcinoma. No control data were reported
(Druckrey et al., 1966).

A single s.c. injection of 50 mg/kg bw (as a 0.8% aqueous solution) produced local sarcomas in 7/15 BD rats, with multiple metastases to the lungs in 3 cases. Tumours appeared between 314 and 740
days. No control data were reported (Druckrey et al., 1970).

(c) Other experimental systems

Intravenous administration: No tumours developed in either of 2
groups of 12 BD rats given weekly i.v. injections of 2 or 4 mg/kg bw
dimethyl sulphate for 800 days (Druckrey et al., 1970), nor in 9 rats
injected with 75-150 mg/kg bw (Swann & Magee, 1968). In the latter
study, no details of the observation period nor of dosing frequency
are available.

Pre-natal exposure: A single i.v. dose of 20 mg/kg bw given to
8 pregnant BD rats on day 15 of gestation induced malignant tumours,
including 3 tumours of the nervous system, in 7/59 offspring observed
for over 1 year. No controls were reported (Druckrey et al., 1970).

3.2 Other relevant biological data

(a) Animals

After an i.v. injection of 75 mg/kg bw in the rat, dimethyl sulphate was no longer detectable in the blood after 3 minutes (Swann, 1968). Dimethyl sulphate is a methylating agent, producing 7-methylguanine in nucleic acids on chemical reaction (Reiner & Zamenhof, 1957), and in vivo in rats (Swann & Magee, 1968).

3.3 Observations in man

Case reports: A 47-year-old worker died from an oat cell carcinoma of the upper bronchus with metastases 11 years after exposure to dimethyl sulphate vapours in a chemical factory. He was reported not to have been a heavy smoker. During the first years of his work with dimethyl sulphate he was in a narrow and badly ventilated room with 6-10 other people; occasional intoxication from dimethyl sulphate was noted in all of them. Further investigations of the other workers showed that 3 had died from bronchial cancer (Druckrey et al., 1966).

4. Comments on Data Reported and Evaluation[1]

4.1 Animal data

Dimethyl sulphate has been shown to be carcinogenic in the rat, the only species tested, by inhalation, subcutaneous injection and following pre-natal exposure. It is carcinogenic to the rat in a single-dose exposure.

4.2 Human data

The case reports mentioned raise some suspicion as to the possible carcinogenicity of dimethyl sulphate in man, but good epidemiological evidence is not available to confirm this.

---

[1] See also the section "Extrapolation from animals to man" in the introduction to this volume.

5. References

Druckrey, H., Kruse, H., Preussmann, R., Ivankovic, S. & Landschütz, Ch. (1970) Cancerogene alkylierende Substanzen. III. Alkylhalogenide, -sulfate, -sulfonate und ringgespannte Heterocyclen. Z. Krebsforsch., 74, 241

Druckrey, H., Preussmann, R., Nashed, N. & Ivankovic, S. (1966) Carcinogene alkylierende Substanzen. I. Dimethylsulfat, carcinogene Wirkung an Ratten und wahrscheinliche Ursache von Berufskrebs. Z. Krebsforsch., 68, 103

Feigl, F. & Goldstein, D. (1957) Spot tests for nitromethane, monochloro (bromo)acetic acid, dimethyl sulphate, iodomethane and methyl sulphuric acid. Analyt. Chem., 29, 1522

Fuchs, J. (1969) Sulfuric and sulfurous esters. In: Kirk, R.E. & Othmer, D.F., eds., Encyclopedia of Chemical Technology, 2nd ed., New York, John Wiley & Sons, Vol. 19, p. 483

Preussmann, R., Schneider, H. & Epple, F. (1969) Untersuchungen zum Nachweis alkylierender Agentien. II. Der Nachweis verschiedener Klassen alkylierender Agentien mit einer Modifikation der Farbreaktion mit 4-(4-Nitrobenzyl)-pyridin (NBP). Arzneimittel-Forsch., 19, 1059

Reiner, B. & Zamenhof, S. (1957) Studies on the chemically reactive groups of deoxyribonucleic acids. J. biol. Chem., 228, 475

Sawicki, E. & Sawicki, C.R. (1969) Analysis of alkylating agents. Application to air pollution. Ann. N.Y. Acad. Sci., 163, 895

Swann, P.F. (1968) The rate of breakdown of methyl methanesulphonate, dimethyl sulphate and N-methyl-N-nitrosourea in the rat. Biochem. J., 110, 49

Swann, P.F. & Magee, P.N. (1968) Nitrosamine-induced carcinogenesis. The alkylation of nucleic acids of the rat by N-methyl-N-nitrosourea, dimethylnitrosamine, dimethyl sulphate and methyl methanesulphonate. Biochem. J., 110, 39

US Tariff Commission (1972) Synthetic organic chemicals, United States production and sales, 1970. TC Publication, 479

DIETHYL SULPHATE*

1. Chemical and Physical Data

1.1 Synonyms and trade names

Chem. Abstr. No.: 64-67-5

Diethyl monosulphate; ethyl sulphate; sulphuric acid, diethyl ester

1.2 Chemical formula and molecular weight

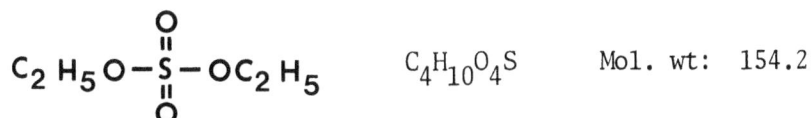

$$C_2H_5O-\overset{\displaystyle O}{\underset{\displaystyle O}{\overset{\|}{\underset{\|}{S}}}}-OC_2H_5 \qquad C_4H_{10}O_4S \qquad \text{Mol. wt: } 154.2$$

1.3 Chemical and physical properties of the pure substance

(a) Description: Colourless, oily liquid

(b) Boiling-point: 209.5°C (with decomposition); 96°C at 15 mm

(c) Melting-point: -25°C

(d) Density: d_4^{25} 1.172

(e) Refractive index: n_D^{18} 1.4010

(f) Solubility: Miscible with alcohol and ether and probably with most polar organic solvents

(g) Volatility: Vapour pressure at 47°C is 1 mm Hg

(h) Stability: Rapidly decomposes in hot water into ethyl hydrogen sulphate and alcohol

(i) Chemical reactivity: An active alkylating agent

---

* Considered by the Working Group in Lyon, June 1973.

1.4 Technical products and impurities

Diethyl sulphate is available as a technical grade product which typically contains 99% of the active ingredient.

2. Production, Use, Occurrence and Analysis

A review on sulphuric and sulphurous esters including diethyl sulphate was published in 1969 (Fuchs, 1969).

(a) Production and use[1]

Diethyl sulphate has been produced commercially for at least 50 years. It can be produced by absorbing ethylene in concentrated sulphuric acid, or by the action of fuming sulphuric acid on ethyl ether or ethyl alcohol. Diethyl sulphate occurs as an intermediate in one method for the production of ethyl alcohol from ethylene.

No information is available on world commercial production. In 1967, two companies were reported to be manufacturing diethyl sulphate in the Federal Republic of Germany, and only one company manufactures this material in the United States. No indications were found that diethyl sulphate is produced in other countries.

Diethyl sulphate is used mainly as an alkylating agent to convert active-hydrogen compounds such as phenols, amines and thiols to the corresponding ethyl derivatives. Since alternative ethylating agents are available for the production of most ethyl derivatives, it is not possible to establish whether they are produced with diethyl sulphate or not. In the case of the quaternary ammonium ethosulphate salts (produced by reaction with tertiary amines), diethyl sulphate must be used as the alkylating agent. US production of six compounds of this type amounted to at least 500 kg each in 1970 (US Tariff Commission, 1972). All six of these compounds were cationic surfactants: (2-aminoethyl) ethyl (hydrogenated tallow alkyl) (2-hydroxyethyl) ammonium ethosulphate; 1-ethyl-2-(8-heptadecenyl)-1-(2-hydroxyethyl)-2-imidazolinium ethosulphate; N-ethyl-N-hexadecyl-morpholinium ethosulphate; N-ethyl-N-(soybean oil alkyl) morpholinium ethosul-

---

[1] Data from Chemical Information Services, Stanford Research Institute, USA.

phate; ethyl dimethyl (mixed alkyl) ammonium ethosulphate; and triethyl oc-
tadecyl ammonium ethosulphate.

In 1966, it was reported that diethyl sulphate could be used as a mutagenic agent to produce a new variety of barley called Luther (Anon, 1966);
however, no evidence was found that it is presently being used commercially
for this purpose.

(b) Occurrence

Diethyl sulphate has not been reported to occur as such in nature.
Although it could occur in the waste streams from plants where it is made
or used, it hydrolyzes rapidly.

(c) Analysis

General methods for the analysis of alkylating agents have been
described (Preussmann et al., 1969; Sawicki & Sawicki, 1969).

3. Biological Data Relevant to the Evaluation
of Carcinogenic Risk to Man

3.1 Carcinogenicity and related studies in animals

(a) Oral administration

Rat: Two groups of 12 BD rats were given weekly doses of 25 or
50 mg/kg bw diethyl sulphate by gavage for 81 weeks (total dose, 1.9
or 3.7 g/kg bw), and the animals were observed until death. In each
group, 1 squamous cell carcinoma of the forestomach was found, and
6/24 rats showed a number of benign papillomas of the forestomach.
No contemporary controls were described (Druckrey et al., 1970).

(b) Subcutaneous and/or intramuscular administration

Rat: Two groups of 12 BD rats were given 1.25 or 2.5% oily solutions of diethyl sulphate s.c. in weekly doses of 25 or 50 mg/kg bw
(total dose, 0.8 or 1.6 g/kg bw) for 49 weeks. Eleven local sarcomas
and 1 glandular carcinoma developed during a mean survival time of
350 days in the 11 surviving rats examined in the 50 mg/kg group.

Two cases of metastasis to the lungs occurred. In 6/12 rats in the 25 mg/kg bw group, 5 sarcomas and 1 myosarcoma developed during an average survival period of 415 days. No contemporary controls were described (Druckrey et al., 1970).

(c) Other experimental systems

Pre-natal exposure: A single s.c. dose of 85 mg/kg bw diethyl sulphate was given to 3 pregnant BD rats on day 15 of gestation. Malignant tumours of the nervous system developed in 2/30 offspring on days 285 and 541. Spontaneous tumours of this type had not been encountered in untreated animals (Druckrey et al., 1970).

3.2 Other relevant biological data

No data are available to the Working Group.

3.3 Observations in man

No data are available to the Working Group.

4. Comments on Data Reported and Evaluation[1]

4.1 Animal data

Diethyl sulphate is carcinogenic in the rat, the only species tested, following subcutaneous administration and pre-natal exposure. There is inconclusive evidence suggesting carcinogenicity in the rat following oral administration.

4.2 Human data

No epidemiological data are available to the Working Group.

---

[1] See also the section "Extrapolation from animals to man" in the introduction to this volume.

5. References

Anon. (1966) Chemical mutation produces new barley variety. <u>Crops and Soil Magazine</u>, <u>12</u>, 28

Druckrey, H., Kruse, H., Preussmann, R., Ivankovic, S. & Landschütz, C. (1970) Cancerogene alkylierende Substanzen. III. Alkyl-halogenide, -sulfate, -sulfonate und ringgespannte Heterocyclen. <u>Z. Krebsforsch.</u>, <u>74</u>, 241

Fuchs, J. (1969) <u>Sulfuric and sulfurous esters</u>. In: Kirk, R.E. & Othmer, D.F., eds., <u>Encyclopedia of Chemical Technology</u>, 2nd ed., New York, John Wiley & Sons, Vol. 19, p. 483

Preussmann, R., Schneider, H. & Epple, F. (1969) Untersuchungen zum Nachweis alkylierender Agentien. II. Der Nachweis verschiedener Klassen alkylierender Agentien mit einer Modifikation der Farbreaktion mit 4-(4-Nitrobenzyl)-pyridin (NBP). <u>Arzneimittel-Forsch.</u>, <u>19</u>, 1050

Sawicki, E. & Sawicki, C.R. (1969) Analysis of alkylating agents. Application to air pollution. <u>Ann. N.Y. Acad. Sci.</u>, <u>163</u>, 895

US Tariff Commission (1972) <u>Synthetic organic chemicals, United States production and sales, 1970</u>. TC Publication, 479

CUMULATIVE INDEX TO IARC MONOGRAPHS ON THE EVALUATION
OF CARCINOGENIC RISK OF CHEMICALS TO MAN

Numbers underlined indicate volume and numbers in italics indicate page.

Cadmium sulphate	2,74
Cadmium sulphide	2,74
Calcium arsenate	2,48
Calcium arsenite	2,48
Calcium chromate	2,100
Carbon tetrachloride	1,53
Chloroform	1,61
Chloromethyl methyl ether	4,239
Chromic oxide	2,100
Chromium	2,100
Chromium dioxide	2,101
Chromium trioxide	2,101
Chrysene	3,159
Cycasin	1,157
o-Dianisidine	4,41
Dibenz(a,h)acridine	3,247
Dibenz(a,j)acridine	3,254
Dibenz(a,h)anthracene	3,178
7H-Dibenzo(c,g)carbazole	3,260
Dibenzo(h,rst)pentaphene	3,197
Dibenzo(a,e)pyrene	3,201
Dibenzo(a,h)pyrene	3,207
Dibenzo(a,i)pyrene	3,215
Dibenzo(a,l)pyrene	3,224
3,3'-Dichlorobenzidine	4,49
1,2-Diethylhydrazine	4,153
Diethyl sulphate	4,277
Dihydrosafrole	1,170
3,3'-Dimethoxybenzidine	4,41
3,3'-Dimethylbenzidine	1,87
1,1-Dimethylhydrazine	4,137
1,2-Dimethylhydrazine	4,145
Dimethyl sulphate	4,271
Haematite	1,29

Hydrazine	4,127
Indeno(1,2,3-cd)pyrene	3,229
Iron-dextran complex	2,161
Iron-dextrin complex	2 161
Iron oxide	1,29
Iron-sorbitol-citric acid complex	2,161
Isonicotinic acid hydrazide	4,159
Isosafrole	1,169
Lead acetate	1,40
Lead arsenate	1,41
Lead carbonate	1,41
Lead chromate	2,101
Lead phosphate	1,42
Lead subacetate	1,40
Magenta	4,57
Maleic hydrazide	4,173
Methylazoxymethanol acetate	1,164
N-Methyl-N,4-dinitrosoaniline	1,141
4,4'-Methylene bis (2-chloroaniline)	4,65
4,4'-Methylene bis (2-methylaniline)	4,73
4,4'-Methylenedianiline	4,79
N-Methyl-N'-nitro-N-nitrosoguanidine	4,183
1-Naphthylamine	4,87
2-Naphthylamine	4,97
Nickel	2,126
Nickel acetate	2,126
Nickel carbonate	2,126
Nickel carbonyl	2,126
Nickelocene	2,126
Nickel oxide	2,126
Nickel subsulphide	2,126
Nickel sulphate	2,127
4-Nitrobiphenyl	4,113

N-(4-(5-nitro-2-furyl)-2-thiazolyl)acetamide	1,*181*
N-Nitroso-di-n-butylamine	4,*197*
N-Nitrosodiethylamine	1,*107*
N-Nitrosodimethylamine	1,*95*
Nitrosoethylurea	1,*135*
Nitrosomethylurea	1,*125*
N-Nitroso-N-methylurethane	4,*211*
Potassium arsenate	2,*48*
Potassium arsenite	2,*49*
Potassium dichromate	2,*101*
1,3-Propane sultone	4,*253*
β-Propiolactone	4,*259*
Saccharated iron oxide	2,*161*
Safrole	1,*169*
Sodium arsenate	2,*49*
Sodium arsenite	2,*49*
Sodium dichromate	2,*102*
Sterigmatocystin	1,*175*
Streptozotocin	4,*221*
Tetraethyllead	2,*150*
Tetramethyllead	2,*150*
o-Tolidine	1,*87*